Singapore Residence

新加坡住宅

佳图文化 编

·广州·

图书在版编目（CIP）数据

新加坡住宅 ：英汉对照 / 佳图文化编. — 广州 ：华南理工大学出版社，2014.8
ISBN 978-7-5623-4305-9

Ⅰ. ①新… Ⅱ. ①佳… Ⅲ. ①住宅 - 建筑设计 - 新加坡 - 英、汉 Ⅳ. ① TU241

中国版本图书馆 CIP 数据核字（2014）第 138758 号

新加坡住宅
佳图文化 编

出 版 人：韩中伟
出版发行：华南理工大学出版社
（广州五山华南理工大学 17 号楼，邮编 510640）
http://www.scutpress.com.cn E-mail: scutc13@scut.edu.cn
营销部电话：020-87113487 87111048（传真）
策划编辑：赖淑华
责任编辑：骆 婷 赖淑华
印 刷 者：广州市中天彩色印刷有限公司
开 本：889mm×1194mm 1/12 印张：27
成品尺寸：285mm×285mm
版 次：2014 年 8 月第 1 版 2014 年 8 月第 1 次印刷
定 价：368.00 元

Preface

Throughout Singapore's architectural history, the creative energy of free will is continuously inspired in the design of residential buildings. What impress people most is the dynamic form and flexibility of the Singapore residences.

The book *Singapore Residence* has carefully selected the latest outstanding residential projects in Singapore, from apartments to single-family houses, from hillside houses to beach houses, trying to present the characteristics and trends of today's Singapore residences. Every selected residence is well interpreted from its planning, design concept, architectural design, etc. With professional drawings, exquisite renderings and high resolution photographs, the book will enable readers to experience the charm of Singapore residences more directly. It will surely be a reference book for the residential architects and designers as well as for the other professionals, to bring them some new inspirations.

前言

纵观新加坡的建筑史，富含自由意志的创意能量在住宅建筑层面不断地迸发，让人尤为惊叹，更让人震撼的是新加坡住宅的动感与灵巧。

《新加坡住宅》精心收录了新加坡最新的优秀住宅项目，从公寓到独栋别墅，从坡地别墅到滨海别墅，力图向读者展示当今新加坡的住宅风貌。全书内容围绕新加坡住宅案例展开，由浅入深，全方位提供案例的整体规划、设计理念、建筑设计等一系列的信息。配合详细的专业技术图以及精美的效果图和实景图，让不同的新加坡住宅形式能够更为直观地展现在读者面前。同时，本书也为广大的住宅设计师和相关专业人士提供实用的参考借鉴，以激发其新的设计灵感。

Contents 目录

Single-family Houses 独栋别墅

Semi-detached Houses 半独立式别墅

Hillside Houses 坡地别墅

Beach Houses 滨海别墅

Apartments 公寓

Single-family Houses
独栋别墅

Elegant Shape　Private Space　Complete Functions

形体雅致　空间私密　功能完备

Keywords 关键词

Novelty and Fashion 新颖时尚

Contrasting Materials 材质反差

Clear Straight Lines 线条明朗

Mimosa Road

Mimosa 路住宅

Location: 6 Mimosa Road, Singapore

Contractor: QS Builders Pte Ltd

Architectural Design: Park + Associates Pte Ltd

Architect in Charge: Lim Koon Park

Design Team: Christina Thean, Tee Teck Hock, Jeeraporn Prongsuriya

Civil & Structural Engineer: C P LIM & PARTNERS

Area: 865 m^2

Photography: Edward Hendricks

项目地点：新加坡 Mimosa 路 6 号

承包商：QS Builders 有限公司

建筑设计：Park + Associates 有限公司

设计负责人：Lim Koon Park

设计团队：Christina Thean, Tee Teck Hock, Jeeraporn Prongsuriya

土建工程： C P LIM & PARTNERS

面积：865 m^2

摄影：Edward Hendricks

Overview

The client commissioned the project outlining a simple brief: double volume living space, 4 bedrooms, reuse as much of the existing structure as possible. The concept and layout of the house evolved and finalized over three meetings with the client. It eventually ended up as a new build.

The site is rectangular and sits at the junction of 2 roads in an established housing estate in the northern part of Singapore. Over the site visit, the architects were enamoured of the burnt orange brick walls of the existing house. These orange bricks wall thus formed the design essence of the house.

Contrasting Design

One enters the house through a large pivot timber door into a single volume gray granite foyer, and through to the timber panelled double volume living space. The intention was to create not just a contrast in volume but also in materials.

Visual Links

In creating a dialogue between indoor and outdoor, all the living spaces on the ground floor are strongly grounded by its relationship to the outdoors. The mahjong room is complemented by a landscape deck with trees; the powder room has a view to a water feature; the living room is fronted by a lotus pond and the dining room is adjacent to a breakfast deck.

The architects consciously tried to create more visual links to the living space as this is the most often used portion of the house within this household. Hence the timber screen corridor with low level seats that line the 2nd storey of the living room and the internal balcony that overlooks the living space from the 2nd storey lounge area. Lastly, the master domain is perceived as a floating steel and glass box resting atop the solid mass of the house, overlooking the surrounding neighborhood.

Residence Materials

The architects aimed to capture modern design through clear straight lines and massive forms compensated by meticulous and creative selection of materials to keep a warm rustic touch to the feel of the house. One of these materials is burnt orange brick that reconnects the history of the original house prominently having exposed bricks all throughout.

6 MIMOSA ROAD

项目概况

根据客户的委托，需要设计双层居住空间和 4 个卧室，并要能最大限度地利用现有的结构。设计团队与客户经过三轮会议交流之后，终于敲定了最后的住宅理念和住宅布局，使其成为一栋全新的建筑。

该矩形住宅位于新加坡北部某居民小区的两条道路的交叉口。经过实地考察之后，设计师们便一眼爱上了房屋既有的焦橙色砖墙，因此它也成为了此次设计的核心。

设计反差

通过一扇大型转轴木门进入房子，再穿过一个铺着灰白色花岗岩的门厅，随即来到了嵌着木板的双层居住空间。这样设计的意图不仅是要营造空间大小的反差，更是要展现材质的反差。

视觉联系

为了能使室内外更好地交流，设计师们将一楼的居住空间与室外环境密切地联系在一起。麻将房外的平台能直通房外的树木林，化妆室里能看到窗外的水景，客厅前方也连着一块荷花池，并且餐厅紧邻露台上的餐桌。

设计师们还一直刻意地去营造更多的与居住空间相关的视觉联系，因为这是与居住在此的家庭最息息相关的事情。因此，放置低矮座椅的木质地板走廊与二楼的客厅连成一线，而在二楼的室内阳台上也能将一楼尽收眼底。此外，该住宅的主体区域被视为一个静止在房子主体上方的可流动的钢铁玻璃结构盒子，俯瞰着周围的一切。

住宅材料

设计师们旨在通过干净明朗的线条、大量的窗户以及经过严谨挑选的富有创造性的材料来保持该住宅温暖田园的感觉，并同时能使住宅造型新颖时尚。这些材料中，住宅大量使用的焦橙色的墙砖又使得大家看到了原住宅的影子。

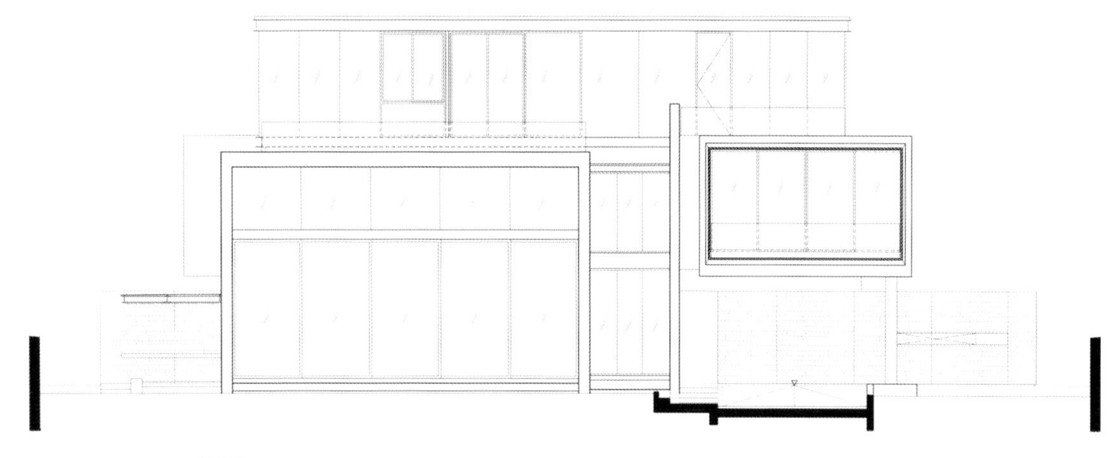

Elevation A　A 立面图

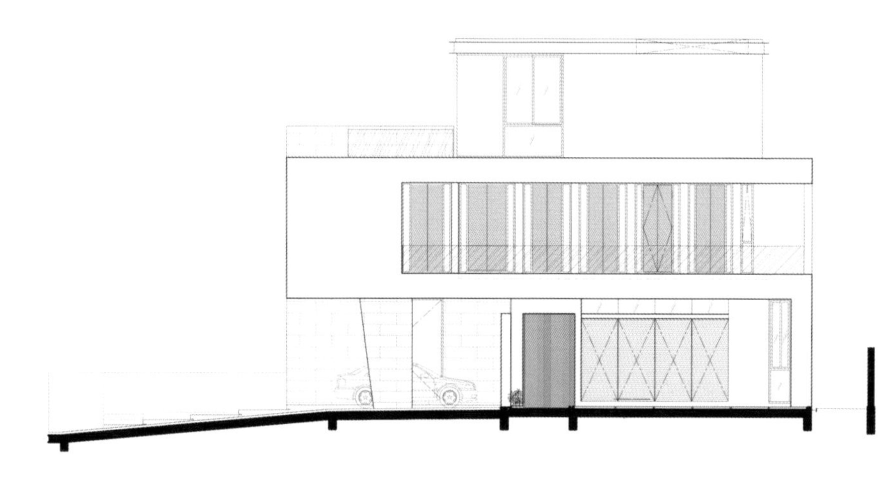

Elevation B　B 立面图

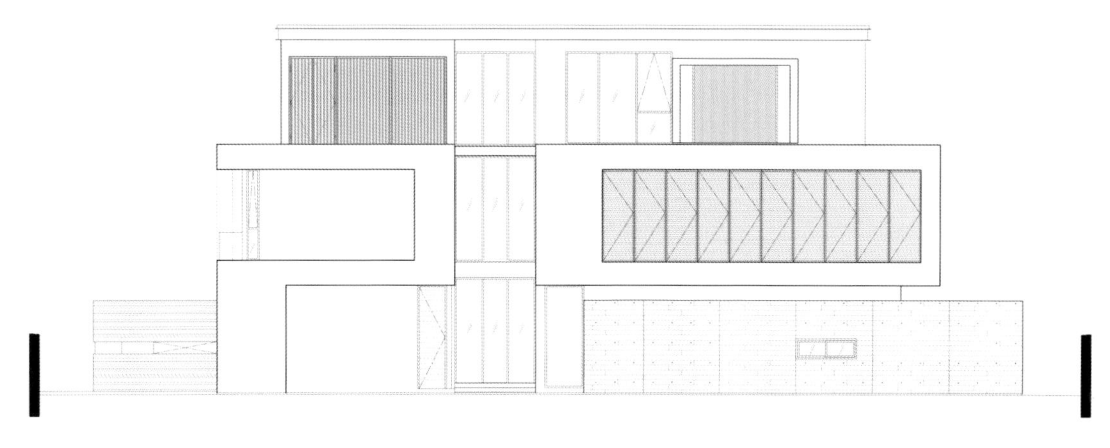

Elevation C　C 立面图

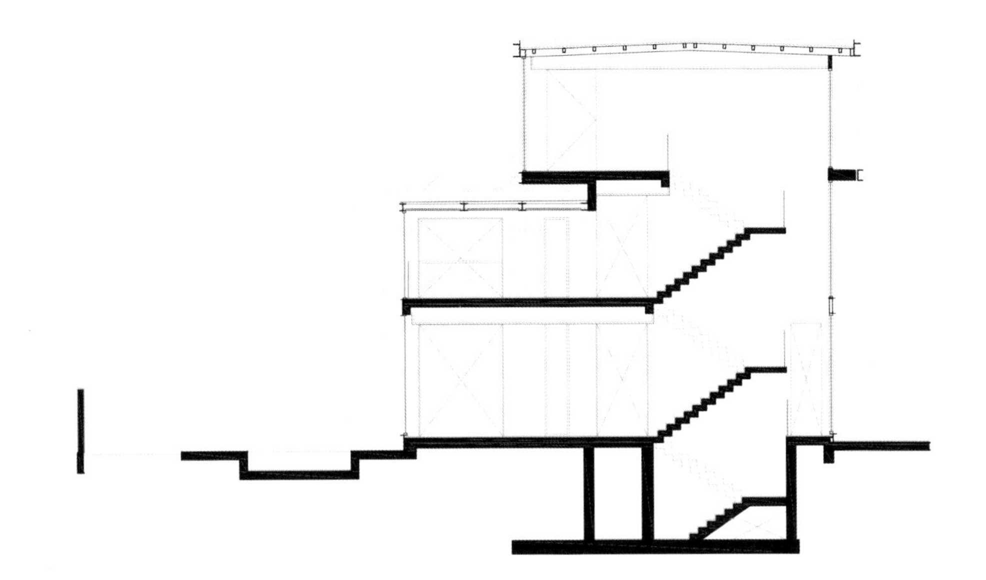

Section 剖面图

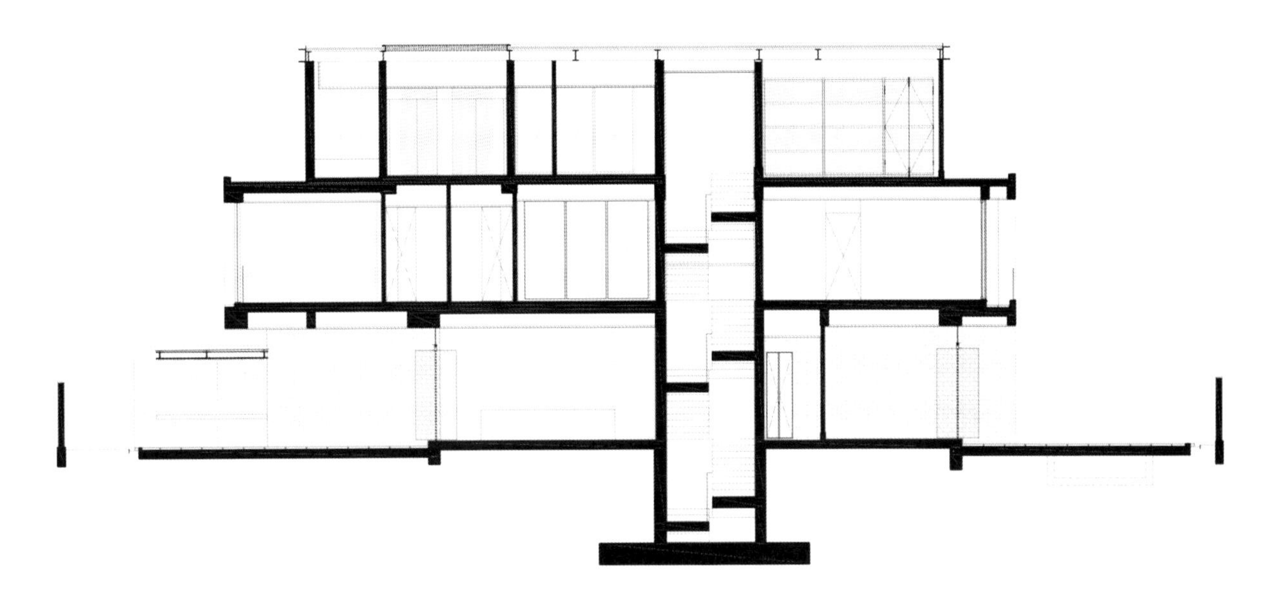

Section 剖面图

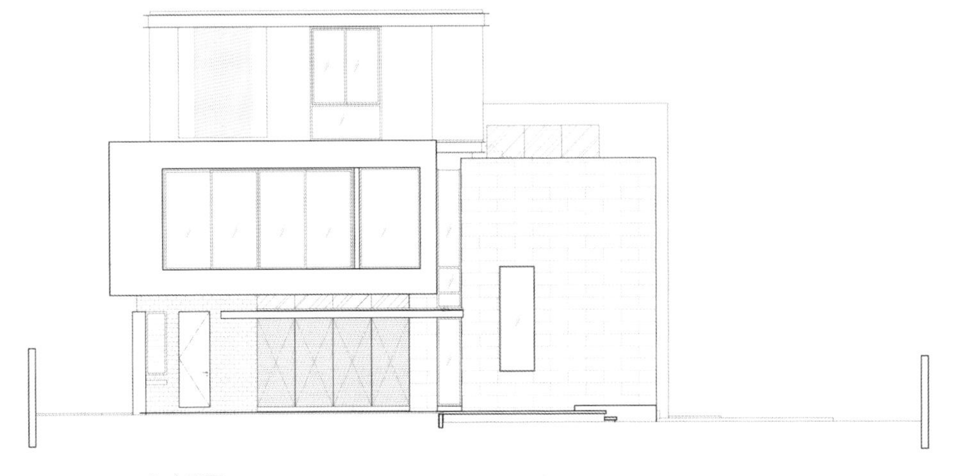

Elevation D　　D 立面图

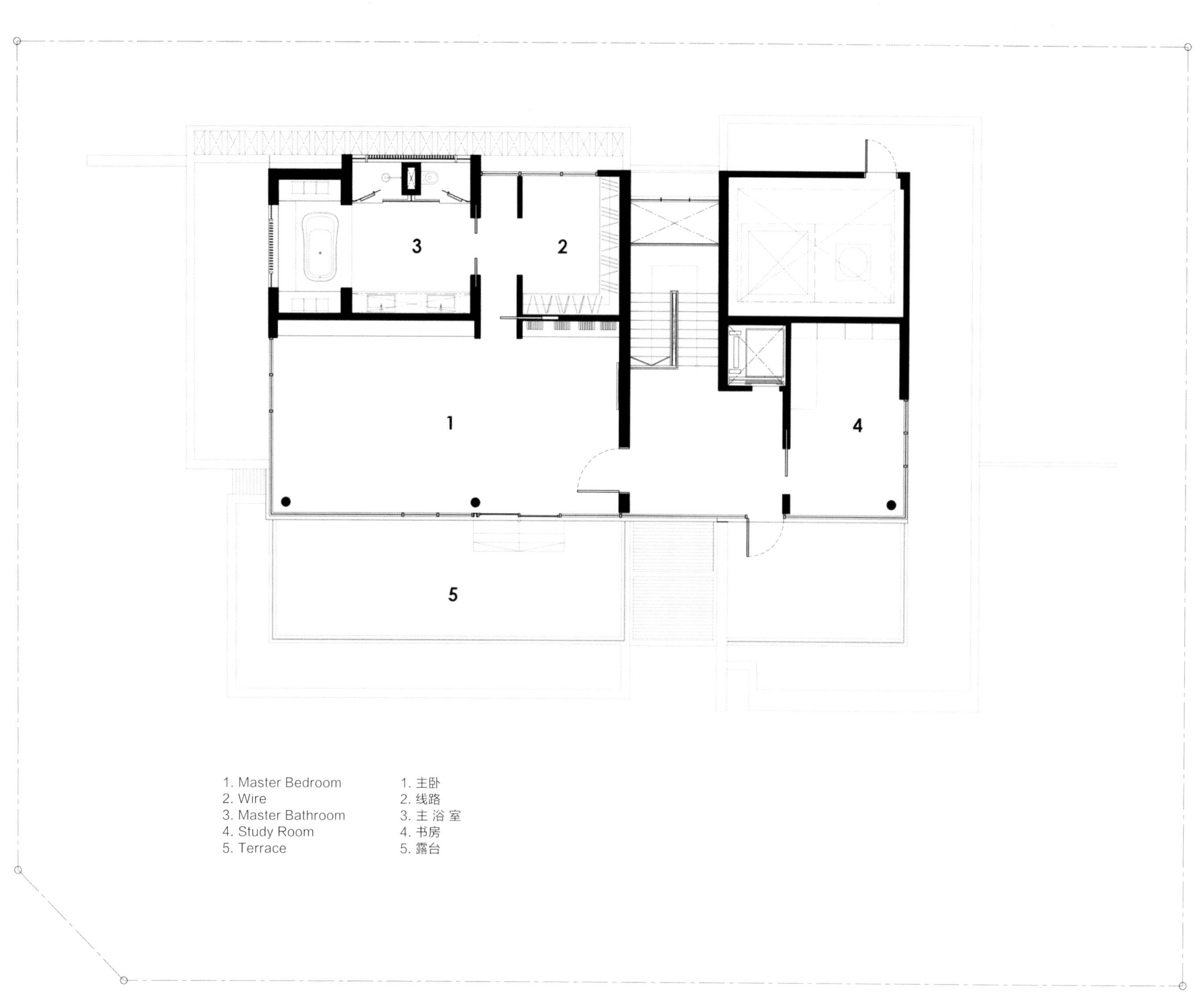

Third Floor Plan　三层平面图

Location: Bukit Timah, Singapore

Architectural Design: TOPOS Design Studio Pte Ltd – Alan Fan, Lim Hong Kian

Main Contractor: Millard Pte Ltd

Quantity Surveyor: Barton Associates Pte Ltd

Civil and Structural Engineer: First Engineer Consultants

Land Area: 950 m^2

Floor Area: 590 m^2

Photography: Derek Swalwell

项目地点：新加坡武吉知马镇

建筑设计：TOPOS 设计事务所——Alan Fan, Lim Hong Kian

总承包商：米勒德私人有限公司

工料测量：巴顿联营私人有限公司

土建工程：第一工程咨询

占地面积：950 m^2

建筑面积：590 m^2

摄影：Derek Swalwell

Keywords 关键词

U-shaped Design U 形设计

Natural and Simplistic 自然简洁

4 Tone Colour Palette 四种色调

Sunset Residence

夕阳住宅

Overview

Situated in a secluded alley in the fashionable area of Bukit Timah in Singapore, this private family residence is a beautiful and understated piece of bespoke and holistic architectural design.

Architectural Design

The modest entrance facade gently invites you into a stunning pool area which reveals the U-shaped plan of the building. This form allows for seclusion as well as views of the pool area from virtually every room in the house as well as fantastic ventilation through full height sliding louver and glass doors. The orientation of the residence make full use of the prevailing wind direction.

The simple no fuss architectural language of the house is further accentuated by a 4 tones to not only highlight the form, but also to allow the client's stunning pieces of furniture to take centre stage. This unpretentious approach in keeping to the natural and simplistic setting of the built environment led to a refined and elegant feel to the spaces, worthy of the esteemed client.

The quality of light and the form on the interior spaces were key to the design which is evident from the generously proportioned lounge and the double height dining area of the first floor. These grand rooms offer fantastic spaces for the family to congregate and enjoy time together.

The second floor of the property is dedicated to the private realms of the users and a relaxing alternative lounge away from the main family area. A comprehensive aluminum louver system, across this floor, aids in sun shading, so as to minimize air-con usage as well as to offer exclusive and spiritual privacy against the surrounding properties.

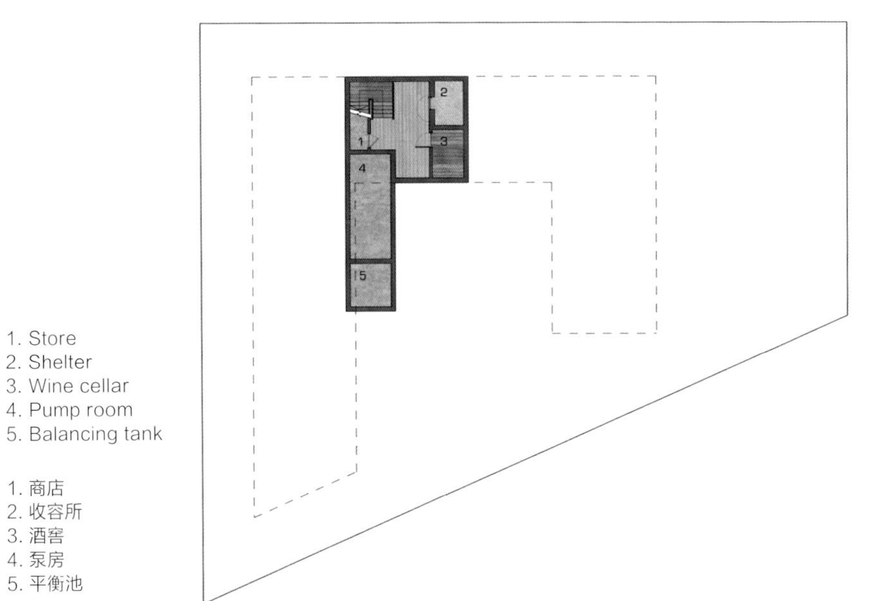

1. Store
2. Shelter
3. Wine cellar
4. Pump room
5. Balancing tank

1. 商店
2. 收容所
3. 酒窖
4. 泵房
5. 平衡池

Basement Plan　地下层平面图

1. Living room
2. Study/Library
3. Dining room
4. Powder room
5. Wet kitchen
6. Dry Kitchen
7. Maid's Room
8. Maid's
9. Music room
10. Master bathroom
11. Walk-in wardrobe
12. Master study
13. Master bedroom
14. Carporch
15. Garden
16. Reflecting pool
17. Patio
18. Swimming Pool
19. Timber decking

1. 起居室
2. 书房
3. 餐厅
4. 盥洗室
5. 湿厨房
6. 干厨房
7. 佣人房
8. 卫生间
9. 音乐室
10. 主卫生间
11. 步入式衣柜
12. 主书房
13. 主卧
14. 停车场
15. 花园
16. 倒影池
17. 露台
18. 游泳池
19. 观景木台

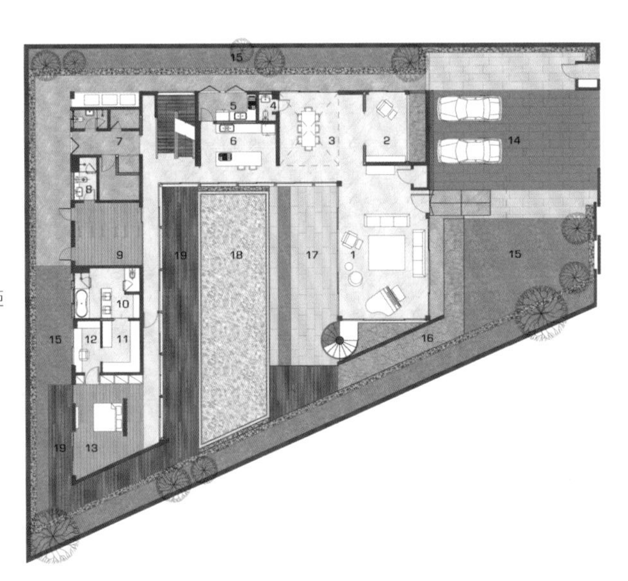

First Floor Plan　一层平面图

项目概况

这座私人家庭住宅坐落在新加坡武吉知马镇时尚地段一个偏僻的小巷子里，是一个美丽朴素的建筑设计。

建筑设计

典雅的入口仿佛绅士邀请来人入内，穿过一个令人炫目的水池，展现了建筑的U形设计。这种设计既隐秘又可以使建筑的每个房间看到水池，落地的滑动百叶窗和玻璃门可形成良好的通风。住宅的朝向充分利用了当地的盛行风向。

房屋设计简洁而不浮夸，四种色调突出了房屋的形态，业主那些令人惊叹的家具占据了中心位置。这种低调的方式保持了自然简洁的建筑环境设定，形成精致优雅的空间感，配得上尊贵的客户。

灯光的质量和室内空间的形式是设计的关键，这一点对于宽敞的休息室和一楼有两层高的餐厅来说尤为明显。这些豪华的房间给家人提供了欢聚一堂的绝佳空间。

住宅的二楼致力于打造成住户的私人空间和轻松的休息室，远离主要的家庭生活区。这一层的综合铝百叶窗系统，可用于遮阴，最大限度降低空调的使用，同时提供独有的精神放松区。

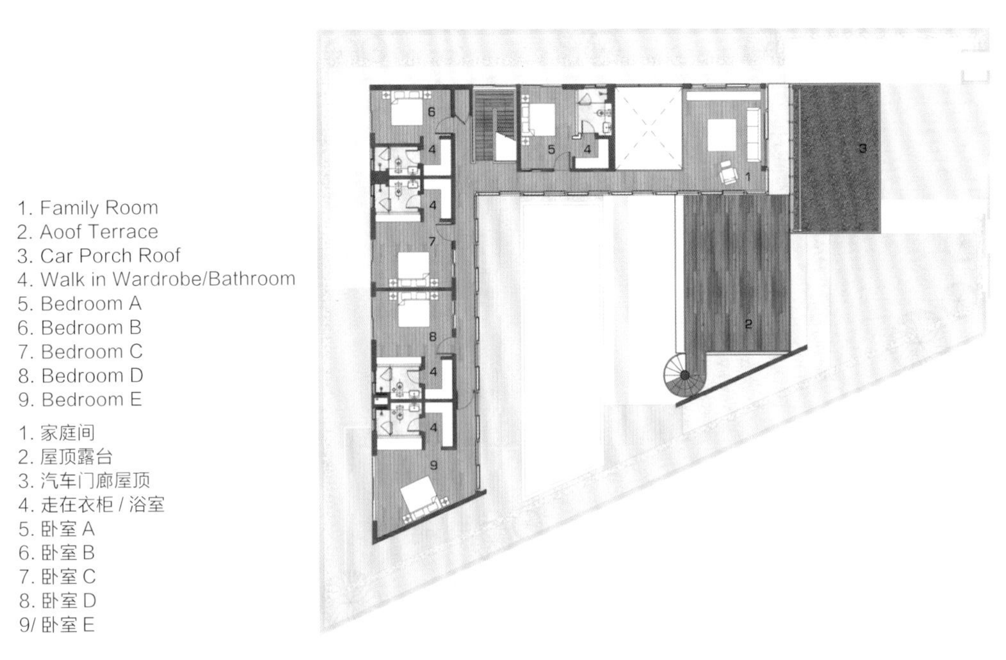

Second Floor Plan 二层平面图

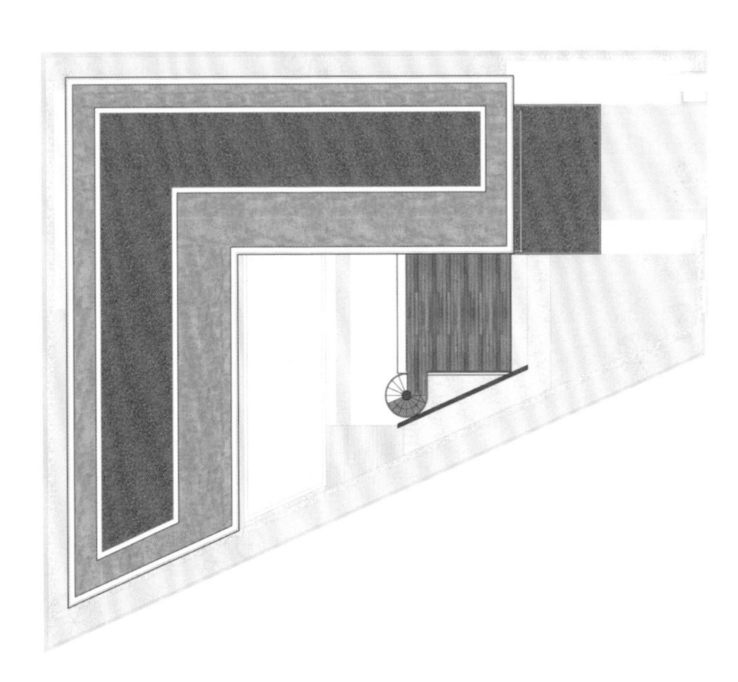

Roof Plan 屋顶层平面图

Location: East Coast, Singapore

Architectural Design: Wallflower Architecture + Design

Design Team: Robin Tan, Cecil Chee & Sean Zheng

Photography: Albert Lim

项目地点：新加坡东海岸

建筑设计：新加坡桂竹香建筑 + 设计

设计团队：Robin Tan, Cecil Chee & Sean Zheng

摄影：Albert Lim

Keywords 关键词

Enclosed House 围合体

Hardwood Louvers 硬木百叶窗

Natural Ventilation 自然通风

Enclosed Open House

围护式的开放别墅

Overview

The owners wanted a spacious, contemporary house that would be as open as possible but without compromising security and privacy at the same time. Surrounded by neighbours on four sides, the solution was a fully fenced compound with a spatial programme that internalised spaces such as pools and gardens, which are normally regarded as external to the envelope of the house. By zoning spaces such as the bedrooms and servants' quarters on alternative levels, i.e. 2nd storey and basement levels, the ground plane was freed from walls that would have been required if public and private programmes were interlaced on the same plane. The see-through volumes allow a continuous, uninterrupted 40-metres view, from the entrance foyer and pool, through the formal living area to the internal garden courtyard and formal dining area in the second volume. All these spaces are perceived to be within the built enclosure of the house.

Architectural Design & Environment Composition

The environmental transparencies at ground level and between courtyards are important in passively cooling the house. All the courtyards have different material finishes (water, grass, granite) and therefore differing heat gain and latency. As long as there are temperature differences between courtyards, the living space, dining area, and pool, house become conduits for breezes that move in between the courtyards, very much like how land and sea breezes are generated. At the second storey, solid hardwood louvers that can be adjusted by hand allow the desired amount of breeze and sunlight to filter through.

Environmentally, the contiguous and interconnected space encourage the slightest breezes, whether they are prevailing and therefore air-movement is horizontal, or convectionally circulated, which the courtyards help to generate. For the owner, it is the experiential serenity that unencumbered space, a gentle breeze, dappled sunlight and the hush of water rippling on a pond that is priceless in our dense and busy urban scape.

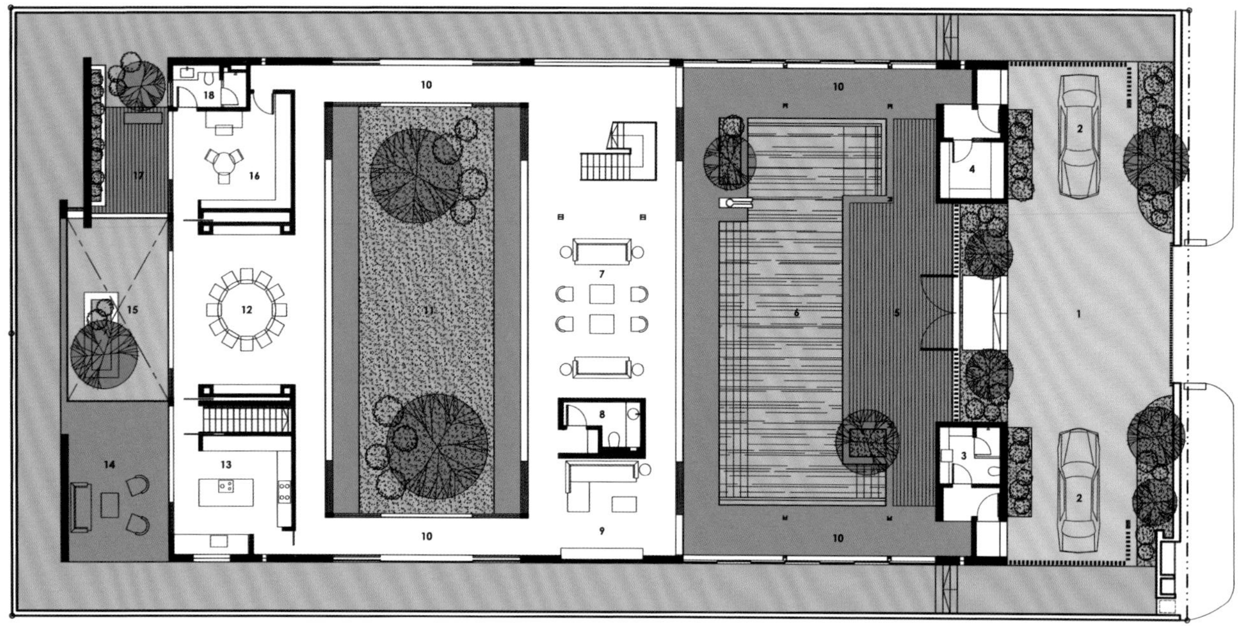

1. Driveway
2. Carporch
3. Changing Room
4. Store
5. Foyer
6. Swimming Pool
7. Living
8. Powder Room
9. TV Area
10. Linkway
11. Courtyard
12. Dining
13. Kitchen
14. Outdoor Terrace
15. Void To Basement
16. Study Room
17. Outdoor Deck
18. Bath

1. 私家车道
2. 车库
3. 更衣室
4. 储藏室
5. 门厅
6. 泳池
7. 客厅
8. 化妆间
9. 电视区
10. 走廊
11. 庭院
12. 餐厅
13. 厨房
14. 露台
15. 地下室入口
16. 书房
17. 露台平台
18. 浴室

Landscaping Layout　景观绿化布局

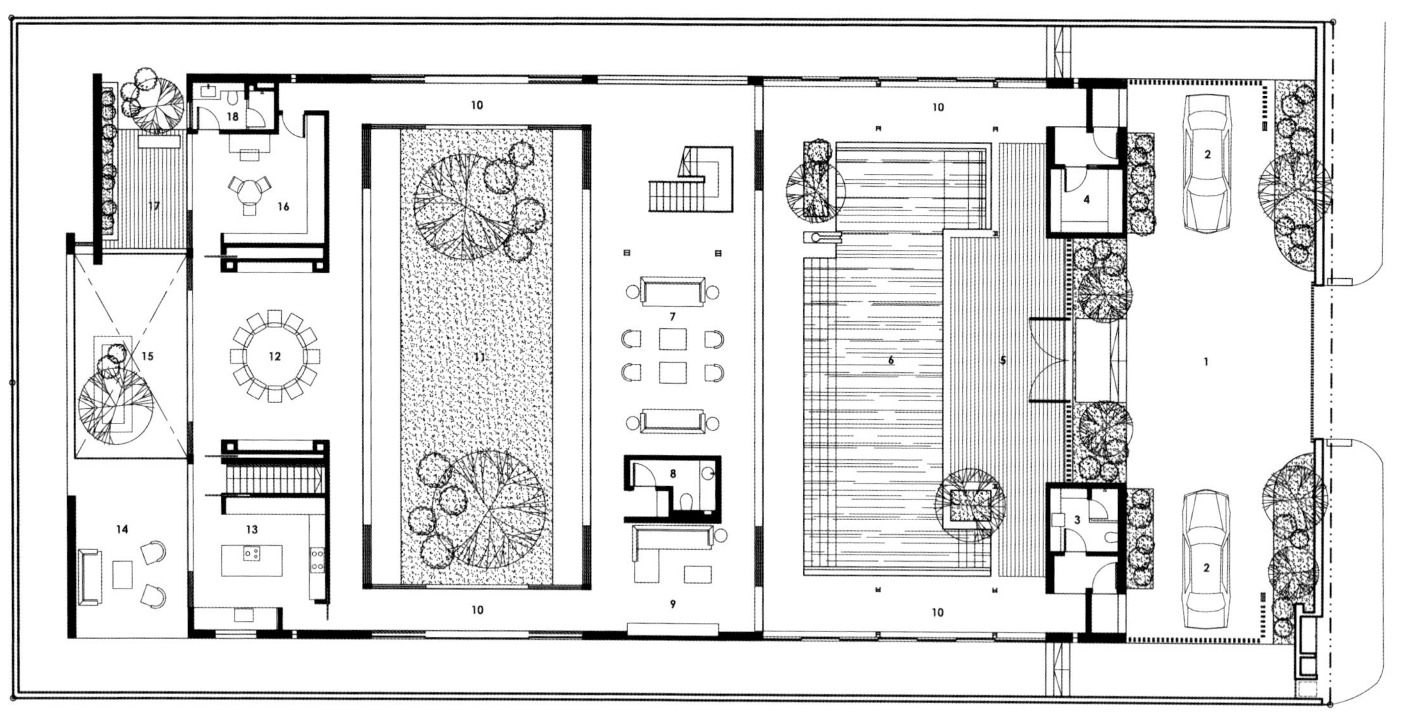

1. Driveway
2. Carporch
3. Changing Room
4. Store
5. Foyer
6. Swimming Pool
7. Living
8. Powder Room
9. TV Area
10. Linkway
11. Courtyard
12. Dining
13. Kitchen
14. Outdoor Terrace
15. Void To Basement
16. Study Room
17. Outdoor Deck
18. Bath

1. 私家车道
2. 车库
3. 更衣室
4. 储藏室
5. 门厅
6. 泳池
7. 客厅
8. 化妆间
9. 电视区
10. 走廊
11. 庭院
12. 餐厅
13. 厨房
14. 露台
15. 地下室入口
16. 书房
17. 露台平台
18. 浴室

First Storey Plan 一层平面图

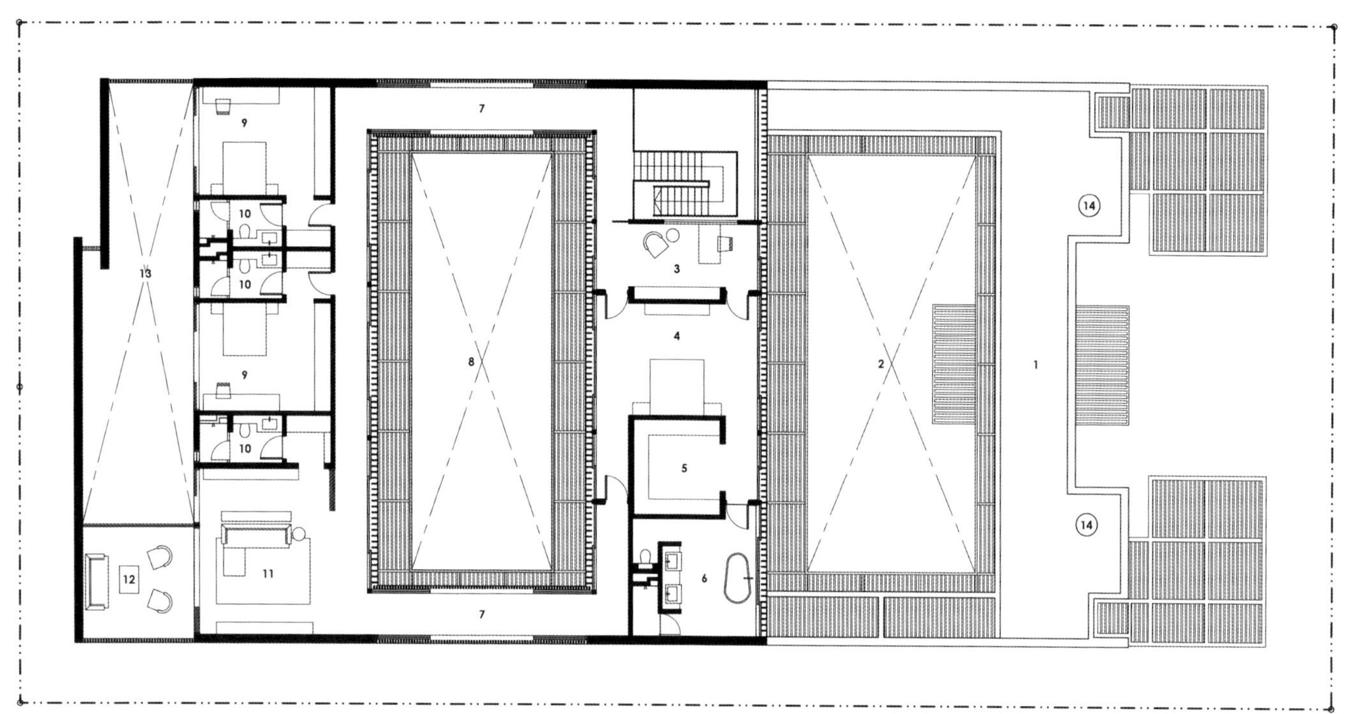

1. Flat Roof
2. Void To Swimming Pool
3. Study
4. Master Bedroom
5. Wardrobe
6. Master Bath
7. Linkway
8. Void To Courtyard
9. Bedroom
10. Bath
11. Family Room
12. Outdoor Balcony
13. Void
14. Skylight

1. 屋顶平台
2. 泳池入口
3. 书房
4. 主卧
5. 衣橱
6. 主浴室
7 走廊
8. 庭院入口
9. 次卧
10 浴室
11. 家庭聚会室
12. 室外阳台
13. 预留空间
14. 天窗

Second Storey Plan 二层平面图

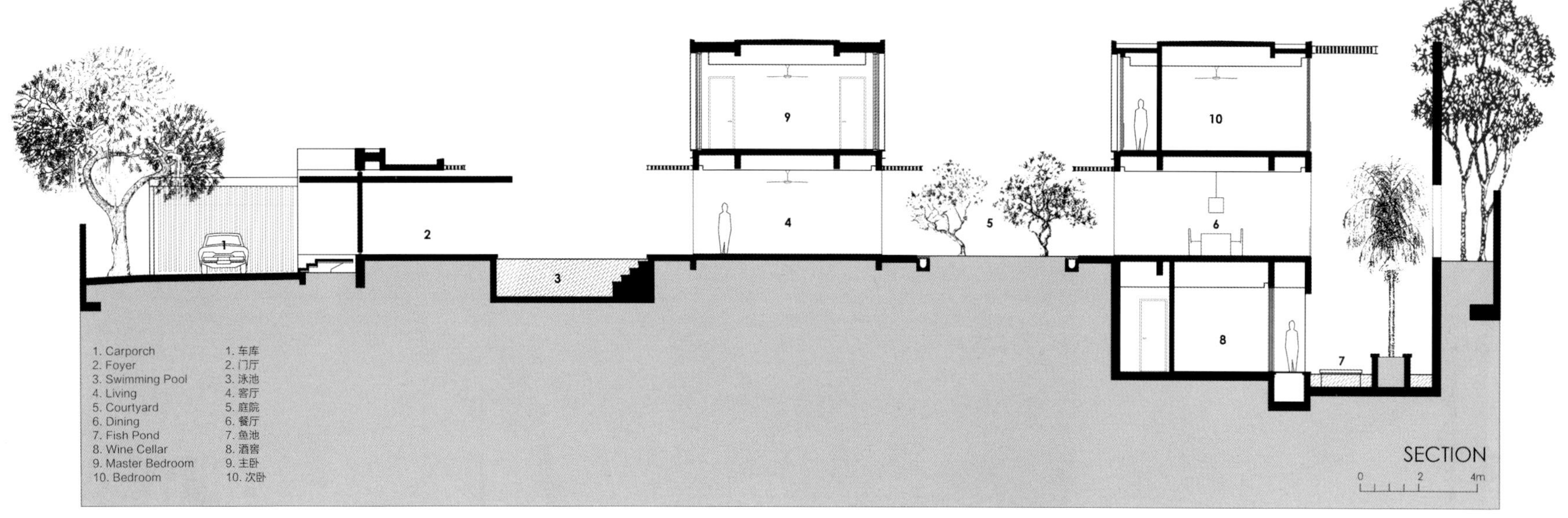

Section 剖面图

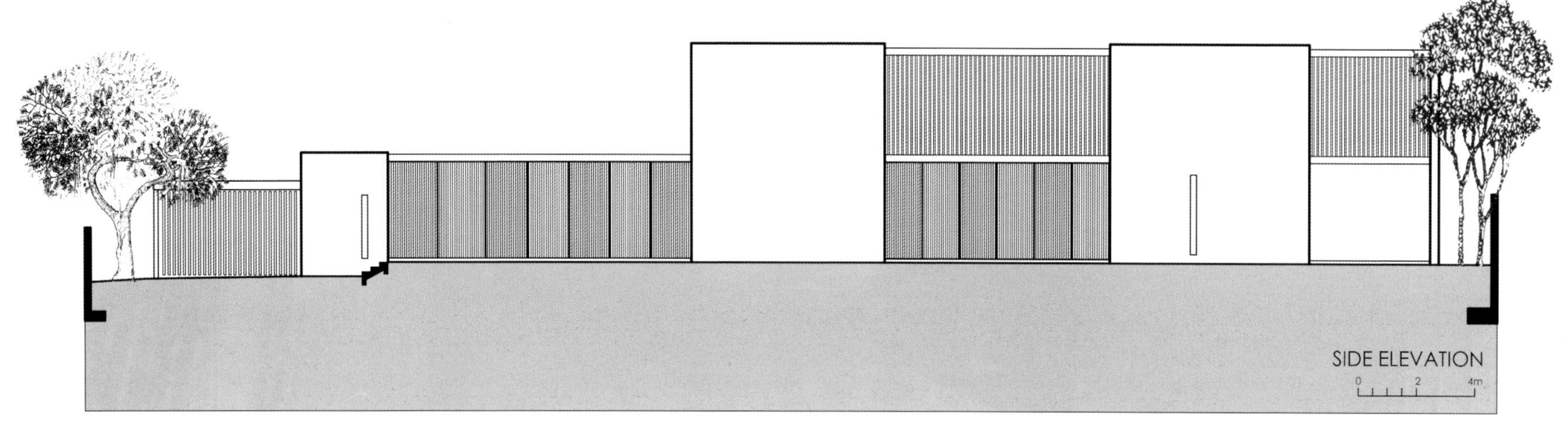

Side Elevation 侧视图

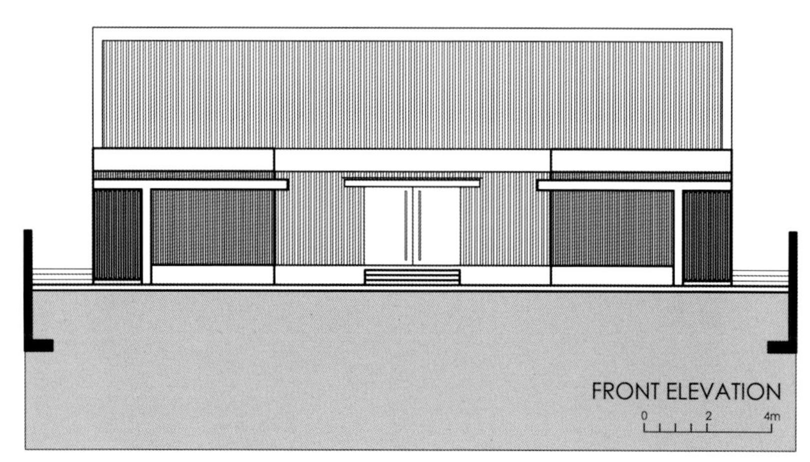

Front Elevation 正立面图

项目概况

业主想要一个宽敞现代的房子，尽可能开放但同时不影响安全和隐私。房子四面都是邻居，解决的方案是用池塘、花园等通常被认为是房屋围护之外的空间组成一个完全合围的混合体。通过空间分区，如卧室、佣人住处位于可选择的层次，而如果公共和私人项目在同一平面上交错进行时是有必要有的。一眼望去拥有 40 m 的视野，可以从进口的门厅和池塘穿过正式的起居区到室内的花园庭院以及第二层正式的用餐区。所有的这些空间都可以在房屋的建筑围护内感知到。

建筑设计与环境构成

在地面上和庭院之间的环境透明度对房屋荫凉十分重要。所有的庭院采用不同的装饰材质（水、草、花岗岩），因此产生不同的热增量和潜在因素。只要庭院、居住空间、餐饮区域、游泳池之间有温度差异，房屋便成为在庭院之间吹动的微风的导管，很像海陆风的形成。在二层，牢固的硬木百叶窗可以手动调整使所需的微风和阳光穿过。

从环境方面讲，无论是不是盛行风，气流运动是不是水平的，或者是不是在庭院的帮助下产生的对流循环，连续的相通的空间都可以形成轻微的风。对于业主来说，不受阻碍的空间、一阵轻柔的微风、斑驳的阳光和池塘上安然荡漾的水都是在这紧凑忙碌的城市景观中一种宁静的体验。

Location: Sentosa Island, Singapore

Architectural and Interior Design: Nicholas Burns Associates

Architectural Design: Nicholas Burns, Desmond Ong, Yonas Kuragi

Photography: Patrick Bingham–Hall

项目地点：新加坡圣淘沙岛

建筑及室内设计：Nicholas Burns 建筑事务所

设计团队：Nicholas Burns, Desmond Ong, Yonas Kuragi

摄影：Patrick Bingham-Hall

Keywords 关键词

Concrete Structure 混凝土结构

Double Glazed Panels 保温玻璃幕墙

Undecorated Materials 简朴材质

Sentosa House

圣淘沙岛住宅

Overview

Located in the tropical climate zone of Singapore, the "Sentosa House" by Australian and Singaporean firm Nicholas Burns Associates utilizes its surrounding environment and spatial logic to create an efficient and timeless home.

Architectural Design

The rectangular site informed a simple footprint with a concrete primary structure and a soft storey on the ground level that contains the garage, driveway and reflecting pond. A concrete core offers internal structural support, providing a central point for mechanical systems, plumbing, vertical circulation, elevator and efficiently services all the peripheral spaces organized around it. The first floor contains two bedrooms with double glazed panels opening views to the lush vegetation outside, and a small gallery facing the street. Above, the master bedroom is oriented towards the back for privacy, with a lounge and study communicating to a full-width balcony on the front facade. Finally, a rooftop terrace gives the users a 360-degree view of the site.

Building Materials

Using only exposed raw materials, such as stone and steel, to minimize the need for unnecessary surface finishes, the exposed concrete shear walls contain a secondary skin of vertical recycled teak planks like louvers that wrap the exterior walls. The screen helps provide a significant aesthetic while shading the internal structure from direct solar gain. The thermal mass of the poured-in-place skeleton helps regulate and stabilize interior temperatures.

1.Driveway　1. 车道
2.Garage　2. 车库
3.Reflection Pond　3. 镜面池塘
4.Lift　4. 电梯
5.Plant　5. 植物
6.Cellar　6. 酒窖
7.Pool Plant　7. 水生植物
8.Store　8. 储藏室
9.Entrance　9. 入口
10.Gallery　10. 走廊
11.Bedroom　11. 卧室
12.Kitchen　12. 厨房
13.Dining　13. 餐厅
14.Living　14. 客厅
15.Laundry　15. 洗衣间
16.Pool　16. 泳池
17.Study　17. 书房
18.Lounge　18. 休息区
19.Master Bedroom　19. 主卧
20.Terrace　20. 露台
21.Balcony　21. 阳台
P.Powder Room　P. 化妆间
B.Bathroom　B. 浴室

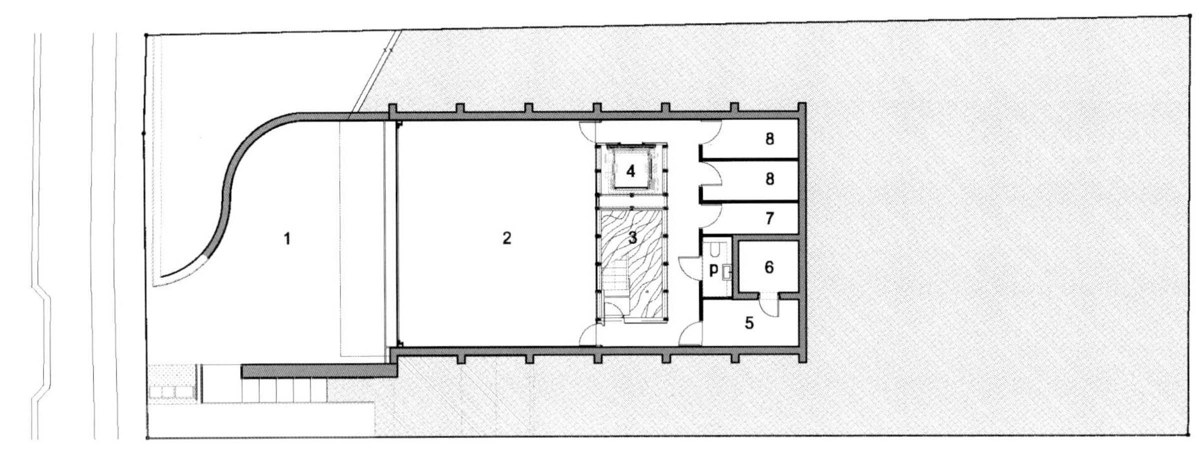

Ground Floor　首层平面图

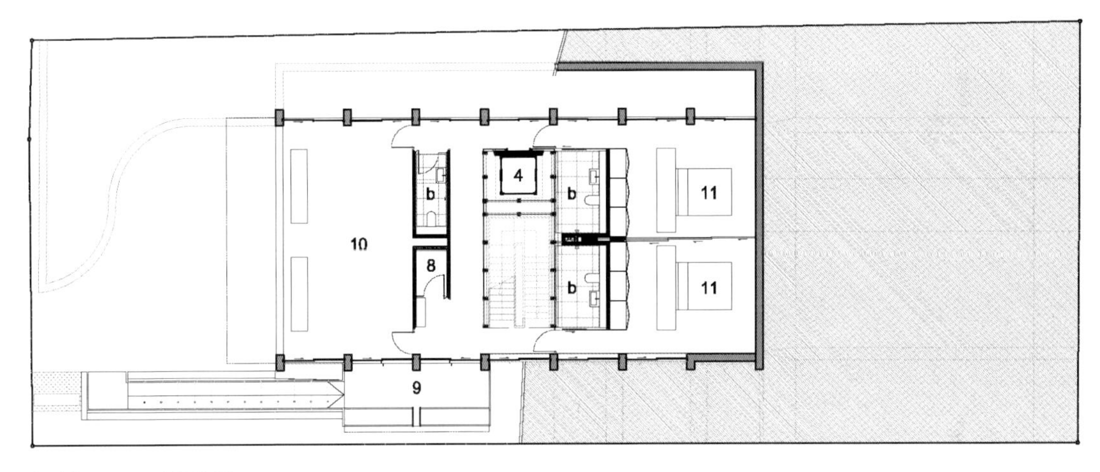

Ist Floor　一层平面图

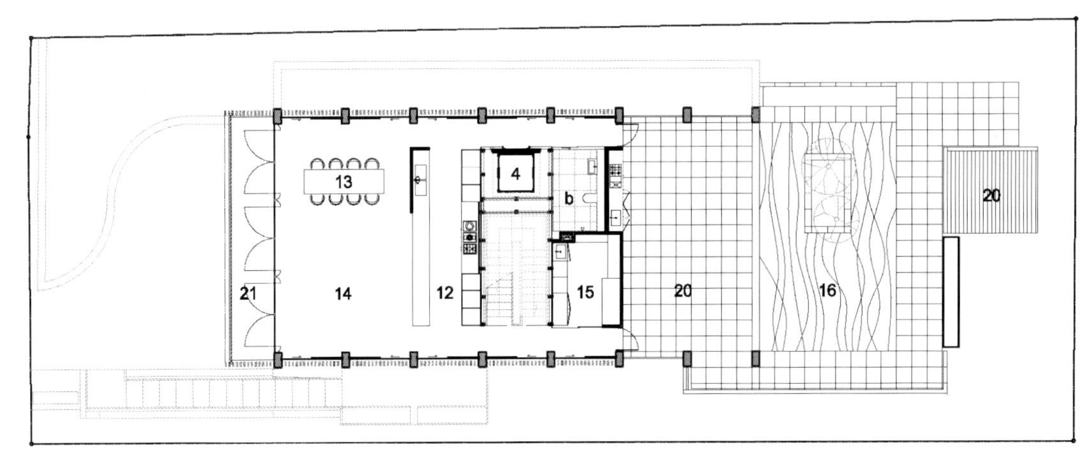

2nd Floor　二层平面图

项目概况

这是由 Nicholas Burns Associates 设计的新加坡圣淘沙岛住宅。设计师利用周围的环境与空间逻辑创造出一座高效而永恒的建筑。

建筑设计

矩形的规整用地形成了简单的平面形状。设计采用了简单的混凝土结构并弱化了地面空间，在此设置了车库、车道和水池。一个混凝土核为结构提供了内部支撑，并在其中设置了机械系统、主水管、垂直交通空间、电梯和高效的服务用房。所有外部空间都围绕着这个核心组织，第一层有两个卧室，双层的保温玻璃幕墙使室内可以观赏到外面丰茂的植物景观。面向街道的是一个小展廊。上层的主卧室朝向后院，休息室和书房位于前院方向，由一个通长的阳台连接起来。此外，在屋顶平台，人们还可以欣赏到周围 360 度的全景景观。

建筑材料

该建筑使用的材料都力求原生态，比如混凝土和钢构件，都未经过过分的精雕细琢。暴露的混凝土剪力墙外还设计了第二层表皮，采用了百叶一样的竖直环保柚木条装饰外表面。这样做不仅美观而富于韵律感，还使混凝土结构免受太阳直射。现场浇筑的保温层有助于维持室内气温的稳定。

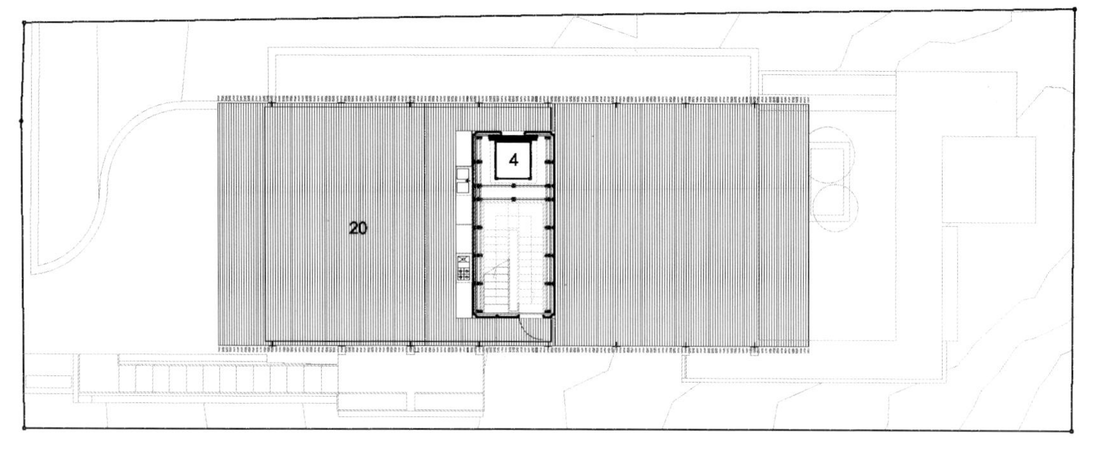

Roof Terrace　屋顶露台

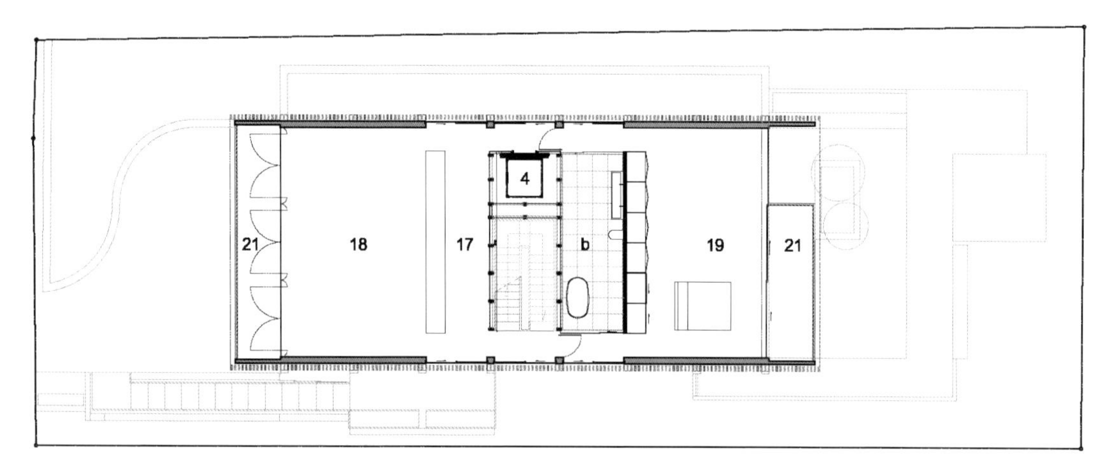

3rd Floor　三层平面图

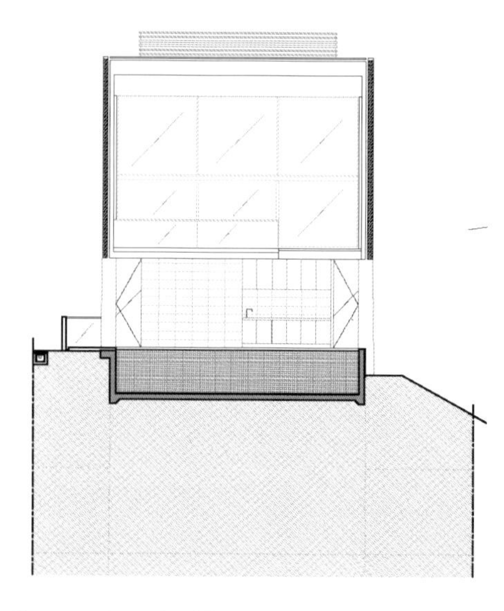

Elevation B　立面图 B

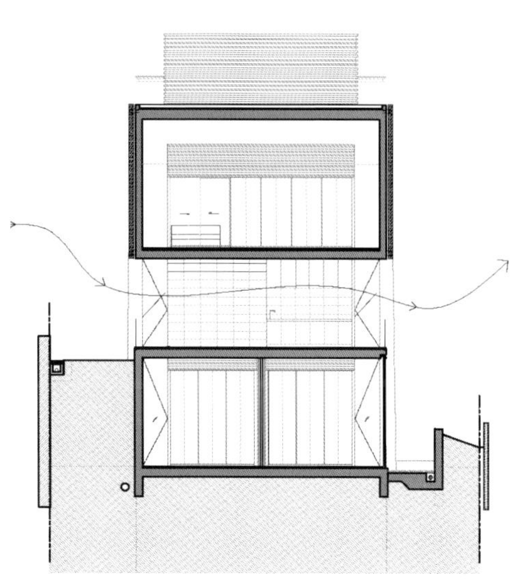

Section　剖面图

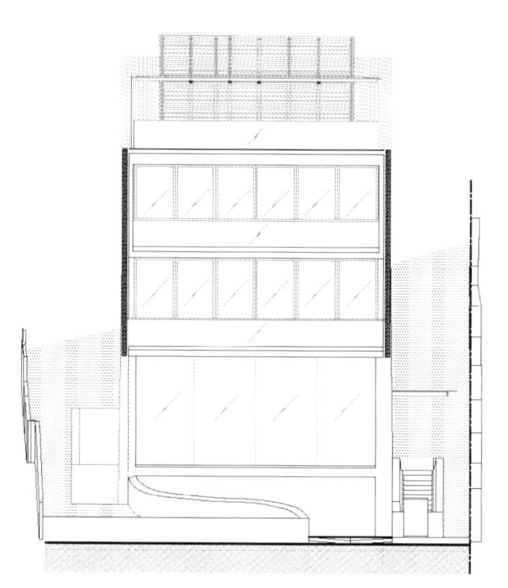

Elevation A　立面图 A

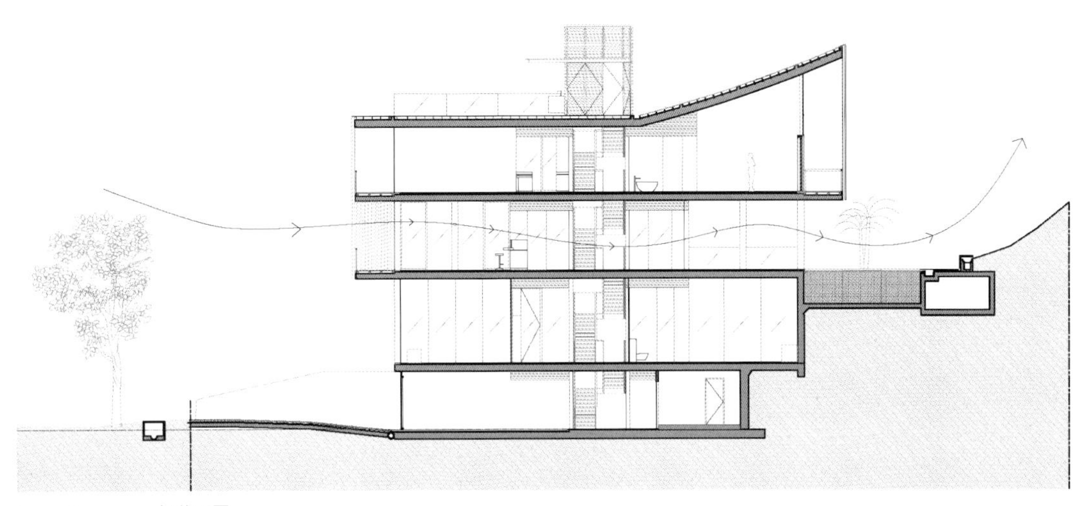

Long Section　长截面图

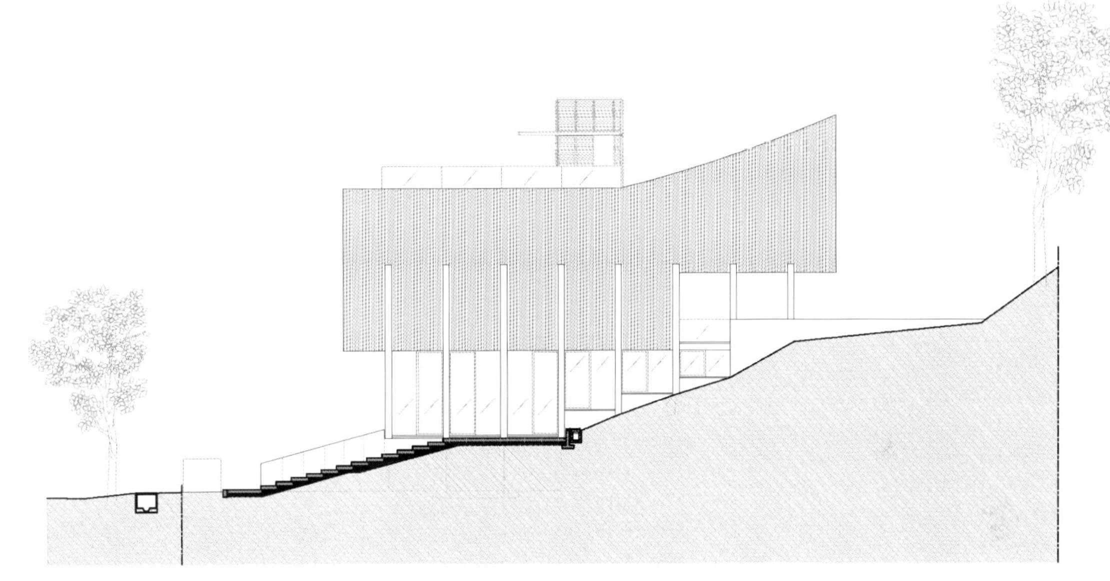

Side-Elevation 1　侧视图 1

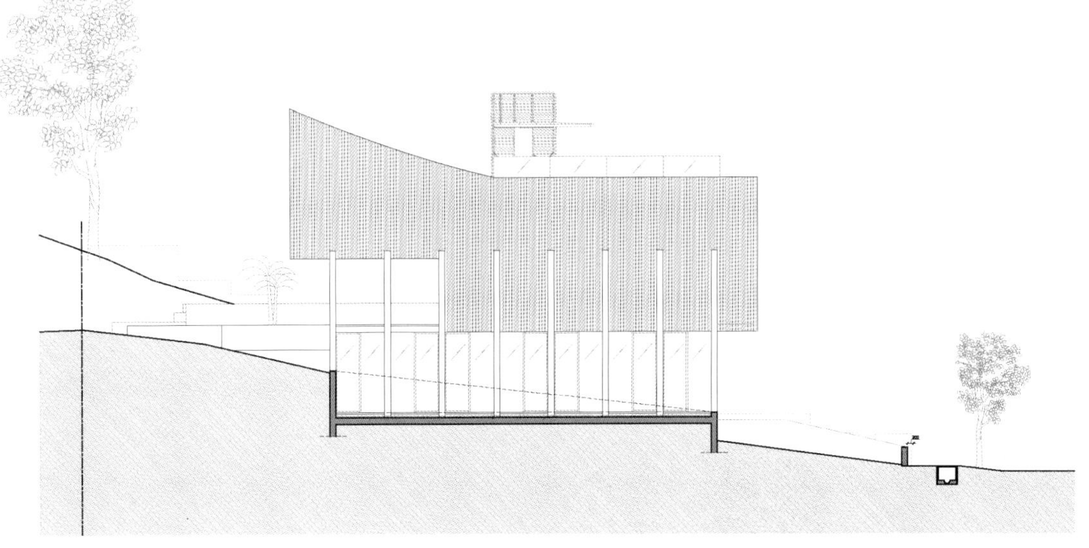

Side-Elevation 2　侧视图 2

Location: Serangoon, Singapore

Architectural Design: Wallflower Architecture + Design

Design Team: Robin Tan, Cecil Chee & Sean Zheng

Land Area: 586 m^2

Floor Area: 558 m^2

Photography: Jeremy San

项目地点：新加坡实龙岗

建筑设计：新加坡桂竹香建筑 + 设计

设计团队：Robin Tan, Cecil Chee & Sean Zheng

占地面积：586 m^2

建筑面积：558 m^2

摄影：Jeremy San

Keywords 关键词

Travertine Wall 石灰华墙壁

Minimal Detail 细节极简

Stone Cladding 石材干挂

Travertine Dream House

梦想之家

Overview and Inspiration

The client's brief for this house was simple. Functionally, to maximize usable area and to incorporate greenery. Aesthetically, to use travertine copiously as an architectural finish. Inspired by the Italian urban-scape during his travels, so too would travertine express this house.

Architectural Composition

The house is organised as two parallel blocks connected by a glass enclosed bridge. The separation between the two blocks allows daylight to stream down to basement spaces. Thick travertine walls and large overhangs are placed on the western side to limit heat gain from the harsh afternoon sun. The entry, living spaces and bedrooms are arranged longitudinally to take advantage of natural cross ventilation and daylight. In order to intensify land use without ending up with an imposing structure, the four storied house has one level sunk into the ground and the other three set away from the access road.

Architectural Design

To accommodate as much green and 'blue' space as possible, the gardens and water bodies are spread throughout the house. The living and dining areas on the ground floor face a swimming pool and a fish pond. The basement's entertainment and guest rooms are open to the sky, with natural light and ventilation coming through a sunken moss garden courtyard. The third storey flat roof is both a recreational deck and a roof garden.

The arrival experience is orchestrated by several layers of travertine wall that suggest a tenuous threshold between the outside and the inside. The detailing is deliberately minimal and precise to enhance the simplicity of the massing and the juxtaposition of solidity and transparency. The narrow blocks that house the living area, the thick stone cladding, multiple levels of gardens and water bodies ensure that the house remains cool in the tropical environment, well ventilated and washed in soft daylight. The three dimensional composition of voids, layers and solids creates spaces for both quiet reflection and family interaction, something for each mood and moment.

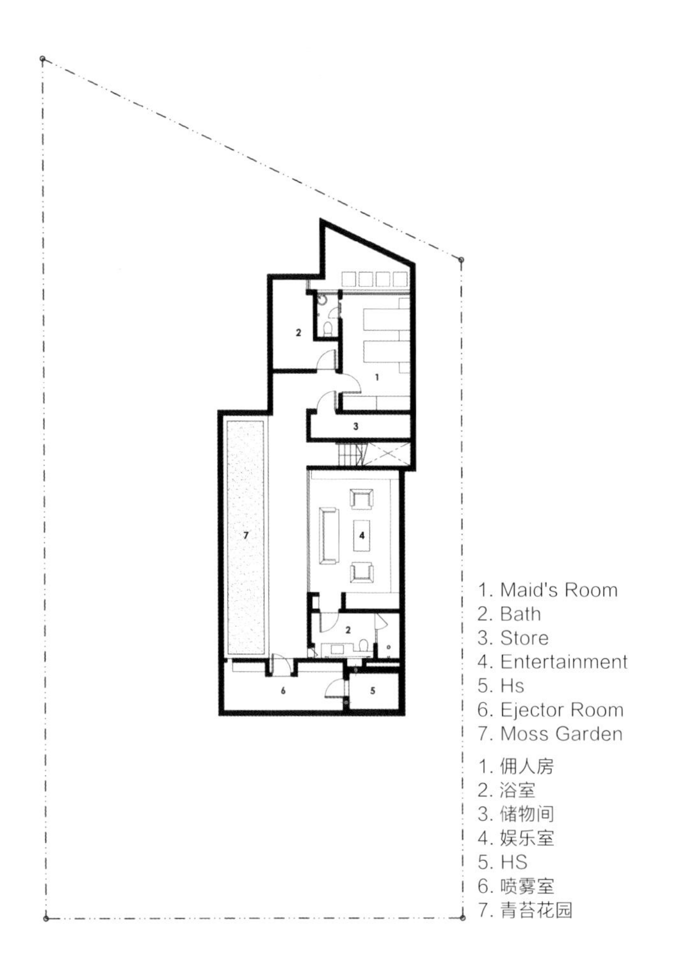

Basement Plan　地下层平面图

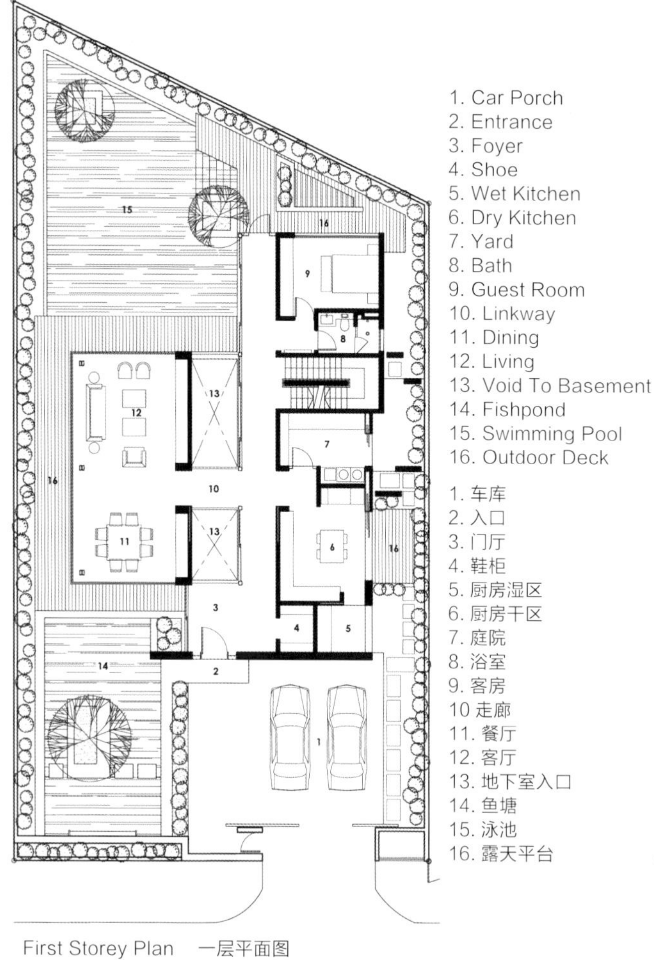

First Storey Plan　一层平面图

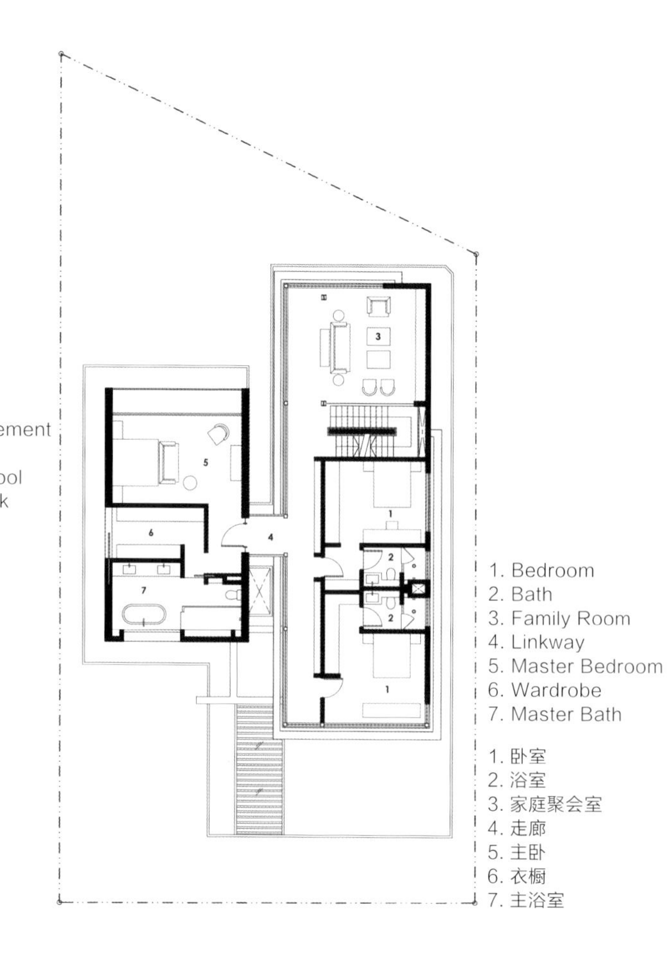

Second Storey Plan　二层平面图

项目概况与设计灵感

这栋豪宅的屋主要求十分简单。在功能性上，要最大限度地提高实用面积，同时更多地容纳绿化。从美学上，要充分利用石灰华作为建筑的外层涂料。这个灵感来自屋主意大利游记中的见闻，同时石灰华也能很好地表达住宅的个性。

建筑构成

豪宅由两个平行结构组成，两者之间用透明的封闭桥梁连接。两个单元的分离能让更多的阳光直接贯穿地下空间。厚厚的石灰华墙壁和巨大的悬挑被放在西侧，以减少过热的太阳光。住宅的入口、起居室和卧室纵向分布，这样能最大限度利用室内的通风和日光。为了高效利用土地，四层的住宅包含了一个地下层和三层内偏向的楼层。

建筑设计

为了容纳更多的绿色和“蓝色”，花园和水体是散布在整个豪宅周边的。在一楼的客厅和餐厅侧面对着游泳池和鱼塘。地下层的娱乐室和客房都是向天空开放，通过下沉式的青苔花园庭院进行自然采光和通风。第三层的屋顶平台成为了休闲区和屋顶花园。

入口处由几层石灰华墙交织而成，显示出室内和室外之间细长的门槛。细节设计刻意极简，并加强了整体的简单性、稳固性和透明性的并存。起居区域狭窄的建筑，厚厚的石材干挂，花园和水体的多层次确保了房屋在热带环境中保持凉爽，拥有良好的通风，沐浴在柔和的日光中。孔洞、层次和固体三维的构成创造了一个既可安静思考又能让家人互动的空间。

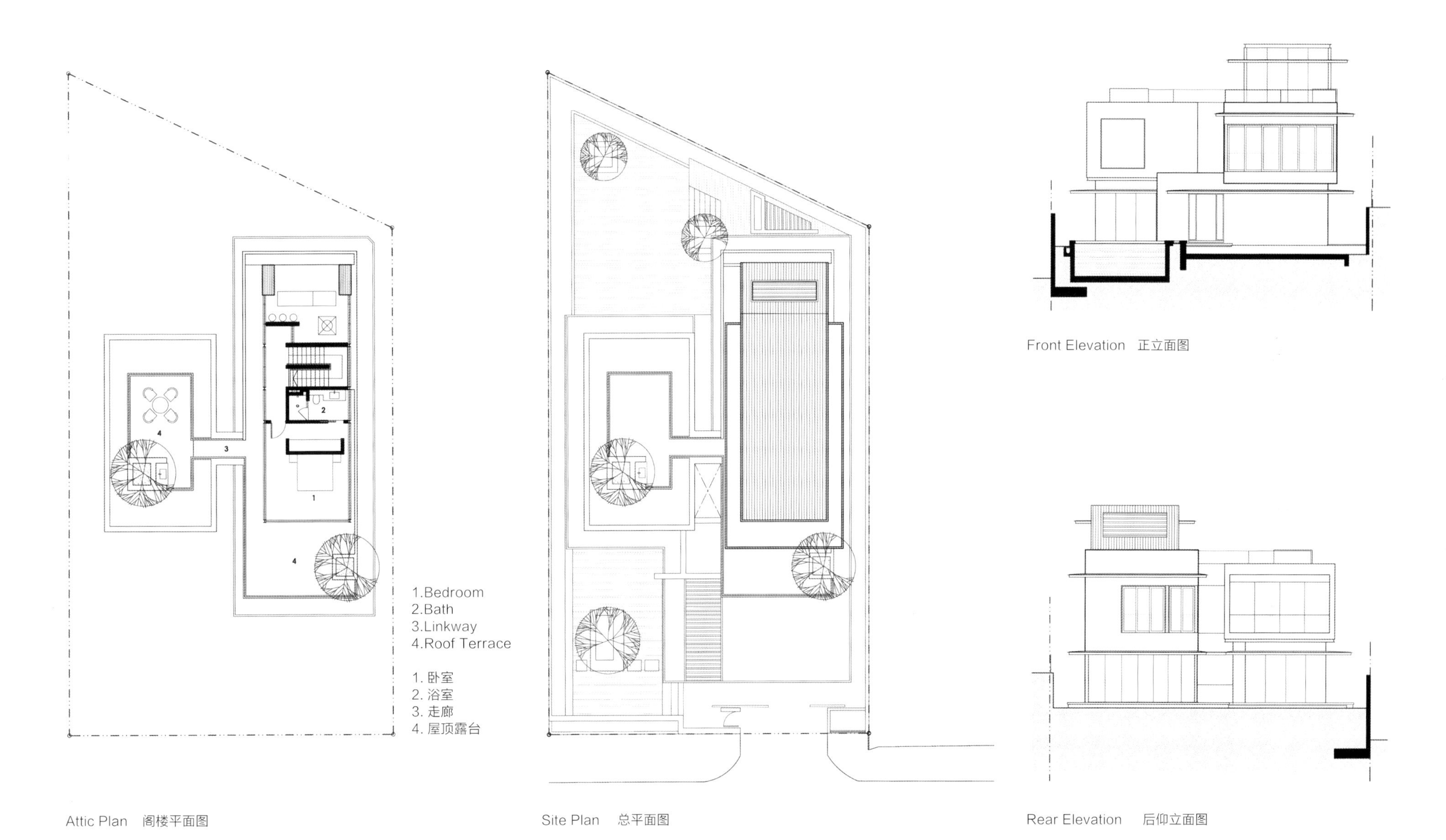

Attic Plan　阁楼平面图

Site Plan　总平面图

Front Elevation　正立面图

Rear Elevation　后仰立面图

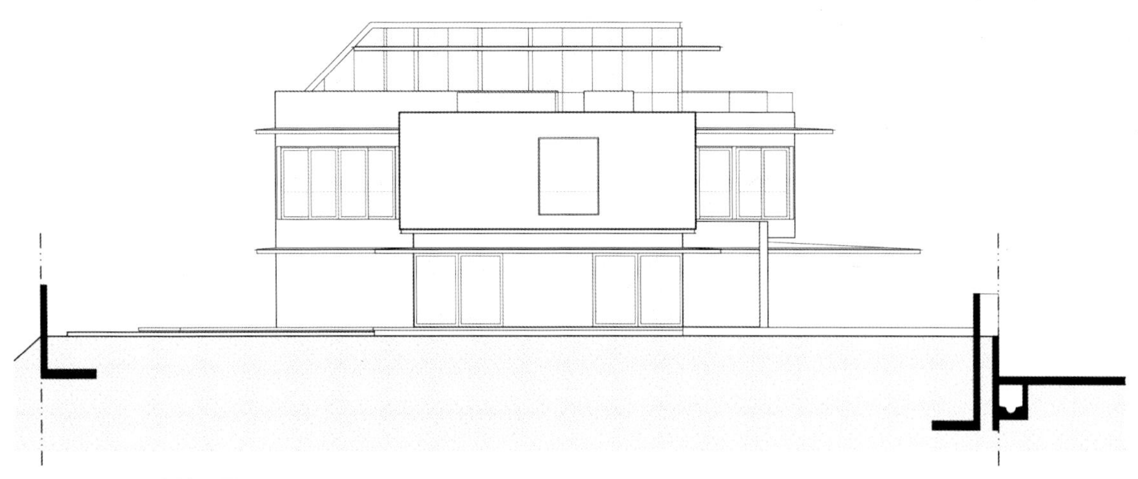

Side Elevation　侧立面图

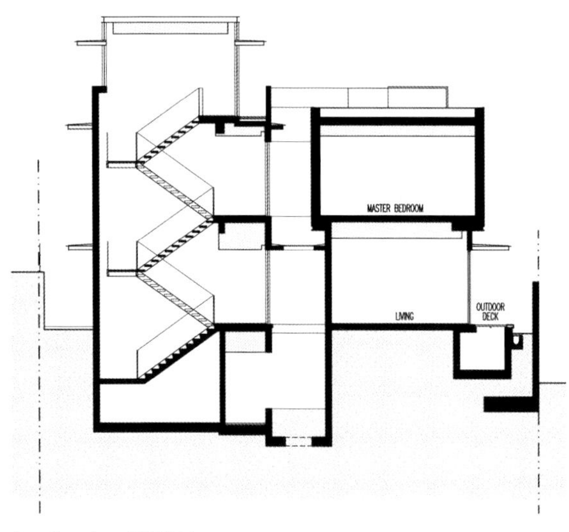

Section A　剖面图 A

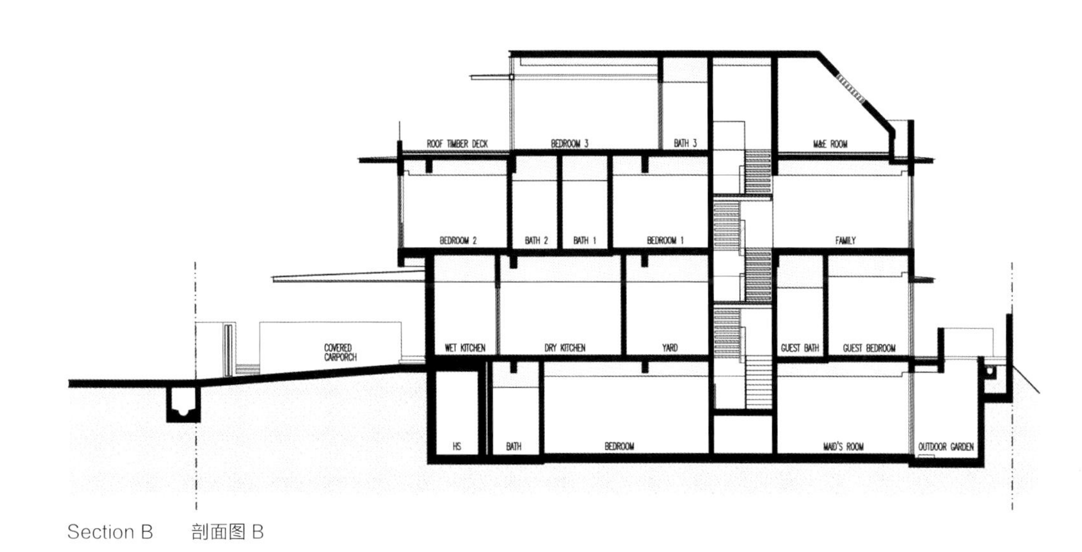

Section B　剖面图 B

Location: Jalan Jintan, Singapore

Architectural Design: Aamer Architects

Land Area: 153 m^2

Floor Area: 236 m^2

Photography : Albert K S Lim

项目地点：新加坡 Jalan Jintan

建筑设计：Aamer 建筑事务所

占地面积：153 m^2

建筑面积：236 m^2

摄影：Albert K S Lim

Keywords 关键词

Rooftop Swimming Pool 屋顶泳池

Spiral Staircase 螺旋楼梯

High Volume Atrium 宽敞中庭

Jalan Jintan

Jalan Jintan 住宅

Overview

A narrow terrace house is transformed into a 3-generational house. Parents, grandma and sister are situated on the ground floor with separate entrances and living areas.

Architectural Design

Common living and dining on the second floor are accessed via grand atrium stairs with sky light through the rooftop pool. Client takes the whole third level while the roof is turned into an outdoor garden with BBQ facilities.

The three storey narrow terrace house manages to accommodate 4 bedrooms with master bedroom on the third level next to a roof top swimming pool. Grand entrance high volume atrium with light from glass bottom pool creates drama over the main stairs up to the 2nd level which is the living/dining area. A caged spiral staircase pops over the atrium as a unique feature of the facade.

项目概况

该项目由一个狭窄的平房改造成一栋可以三世同堂的房子。主人的父母、外婆和妹妹住在一楼，各自拥有独立的入口和活动空间。

建筑设计

房子中庭的楼梯可尽享屋顶泳池透过来的光线，直通二楼的客厅和餐厅。三楼由主人独有，楼顶建有户外花园并配有烧烤装置。

这栋三层高的房子旨在容纳位于一楼的 3 间卧室以及三楼紧邻屋顶泳池的主人卧室。从房子入口处宽敞的中庭拾级而上，通往二楼的楼梯在屋顶泳池天窗的光线照射下，营造出独特的效果。并且，从中庭延伸上来的封闭式螺旋楼梯使得该建筑新颖独特。

34

1. Bedroom 1
2. Bathroom 1
3. Bedroom 2
4. Bathroom 2
5. Bedroom 3
6. Bathroom 3
7. Family room
8. Terrace
9. Carporch
10. Entrance Foyer

1. 卧室 1
2. 浴室
3. 卧室 2
4. 浴室 2
5. 卧室 3
6. 浴室 3
7. 家庭聚会室
8. 露台
9. 车库
10. 入口门厅

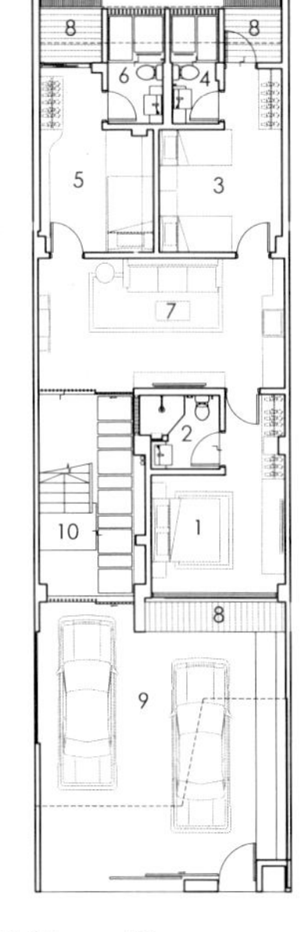

1St Storey Plan

一层平面图

1. Foyer
2. Void
3. Planter
4. Balcony
5. Powder Room
6. Kitchen
7. Dining
8. Livlng

1. 门厅
2. 空地
3. 植物
4. 阳台
5. 化妆间
6. 厨房
7. 餐厅
8. 客厅

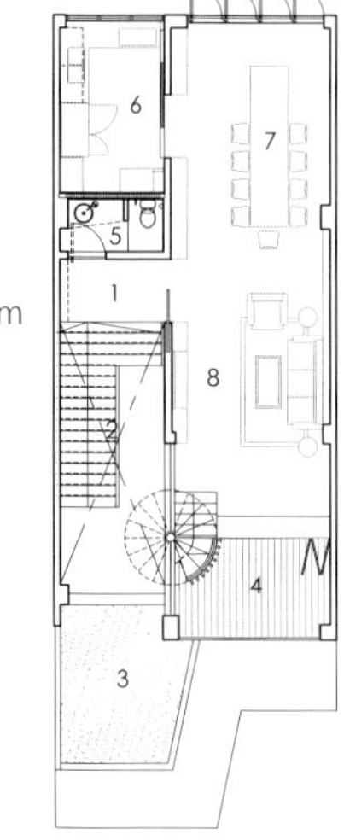

2nd Storey Plan

二层平面图

1. Swimming Pool
2. Balcony
3. Pool Deck
4. Master Bedroom
5. Master Bathroom
6. Walk-In Wardrobe

1. 泳池
2. 阳台
3. 泳池甲板
4. 主卧
5. 主卧浴室
6. 衣帽间

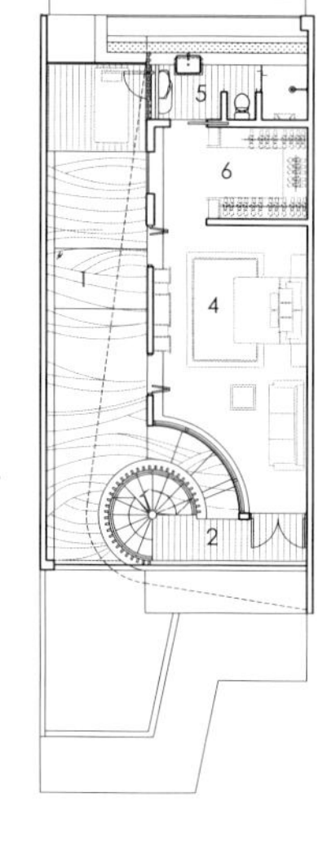

3rd Storey Plan

三层平面图

1. Roof Garden
2. Flat Roof
3. M&E

1. 屋顶花园
2. 平顶
3. 机电设备

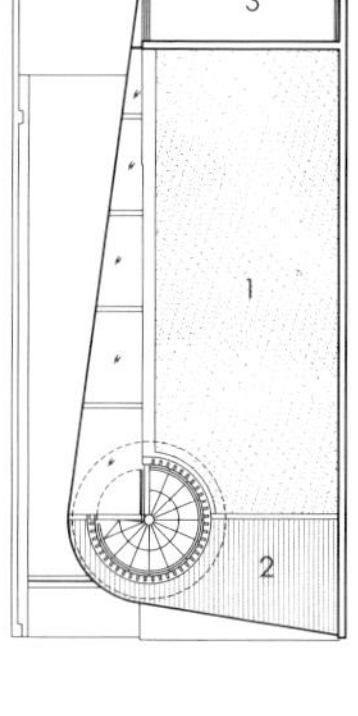

Roof Terrace Storey Plan

屋顶层平面图

Location: East Coast, Singapore

Architectural Design: Wallflower Architecture + Design

Design Team: Robin Tan, Cecil Chee, Sean Zheng, Shirley Tan & Eileen Kok & Roslindah

Land Area: 781 m^2

Floor Area: 796 m^2

Photography: Albert Lim

项目地点： 新加坡东海岸公园大道

建筑设计：新加坡桂竹香建筑 + 设计

设计团队：Robin Tan, Cecil Chee, Sean Zheng, Shirley Tan & Eileen Kok & Roslindah

占地面积：781 m^2

建筑面积：796 m^2

摄影：Albert Lim

Keywords 关键词

Timber Screens 木质屏风

Central Court 中央庭院

Enclosure 围合造型

Centennial Tree House

百年树屋

The owner wanted external blank walls, fixed screens, centre courtyard for light and air. These summed up for them, the tangible facets of an ideal home, a protective enclosure of solitude.

That fortitude and strength is visually given expression by a hundred year old frangipani tree literally found within, centred in a large grassed courtyard surrounded with water. The tree was given a new lease of life having been rescued from a Holland Road site slated for new development.

True to the owners' requirements, the facade is entirely sealed off in most areas, and veiled by fixed timber screening in others. The purity of intention to subjectivity results in a purity of architectural elevation on three sides; there is no yard, opening, back of house, but a pebbled path between a rhythmic timber screen and a lush wall of polyalthia. Visually, the aesthetics exclude both physically and psychologically, but the timber screens along the periphery of the 1st storey allow breezes to come through, refreshing the sheltered corridors and living spaces. The central court encourages this, acting as both a light and air well. Throughout the day as the environment changes, The only area where the timber screens can be opened is between the second storey master bedroom and the court. Motors silently fold the screens away, linking the court to the bedroom.

业主想要空白的外墙和固定屏幕以及光线和空气充足的中心庭院。概括来说，就是一个有形的理想家园，一个孤独的保护壳。

一颗已经上百年的老素馨树，伫立在庭院的草坪中央，四周环水，从视觉上呈现出刚毅和力量的感觉。从即将要有新发展的荷兰路中解救出来，这棵老树获得了新生。

尊重业主的要求，建筑外观的大多数区域完全封锁，其余区域则用固定的木材板覆盖。纯粹的主观意向导致建筑只有纯粹的三面立面；没有院子、开口和后院，但在整齐的木材屏幕和苍翠繁茂的暗罗属植物墙之间有一道鹅卵石小径。视觉上来看，微风沿着一层外围的木材屏幕习习而过，让遮阴的走廊和生活空间重新变得清爽。中央庭院充当了这种光线充足的通风井。环境整天都有变化。木材屏幕唯一能打开的地方在第二层的主卧和庭院之间。屏幕自动折叠，连通庭院和卧室。

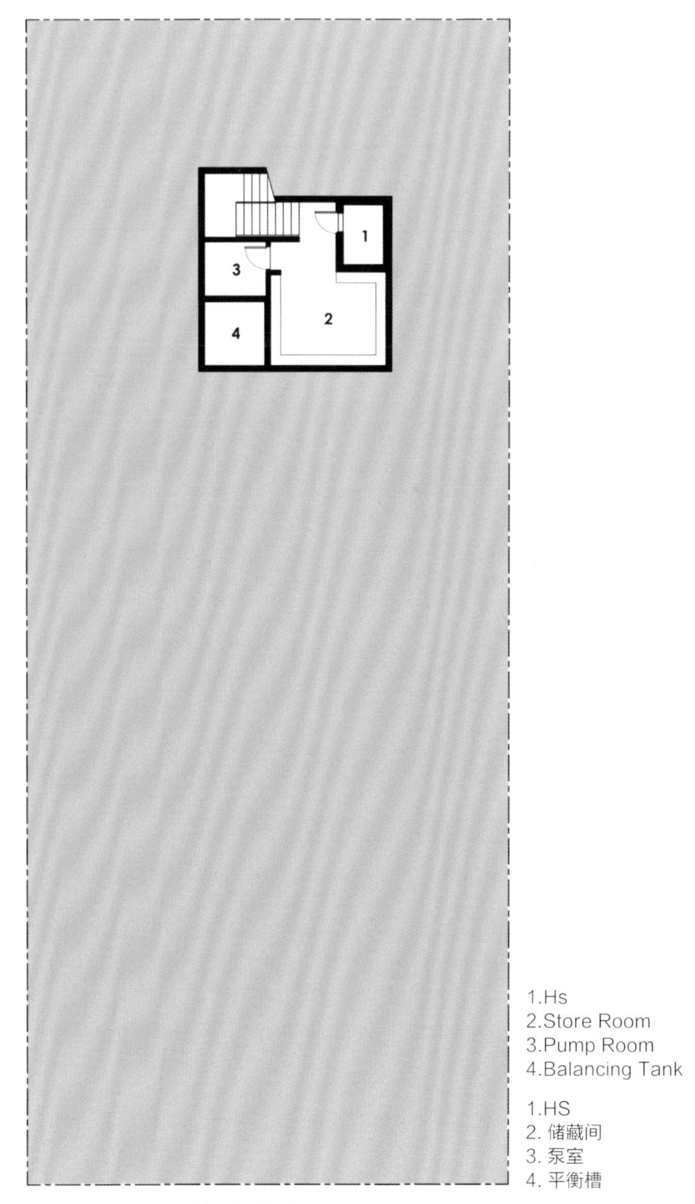

1.Hs
2.Store Room
3.Pump Room
4.Balancing Tank

1.HS
2. 储藏间
3. 泵室
4. 平衡槽

Basement Plan　地下层平面图

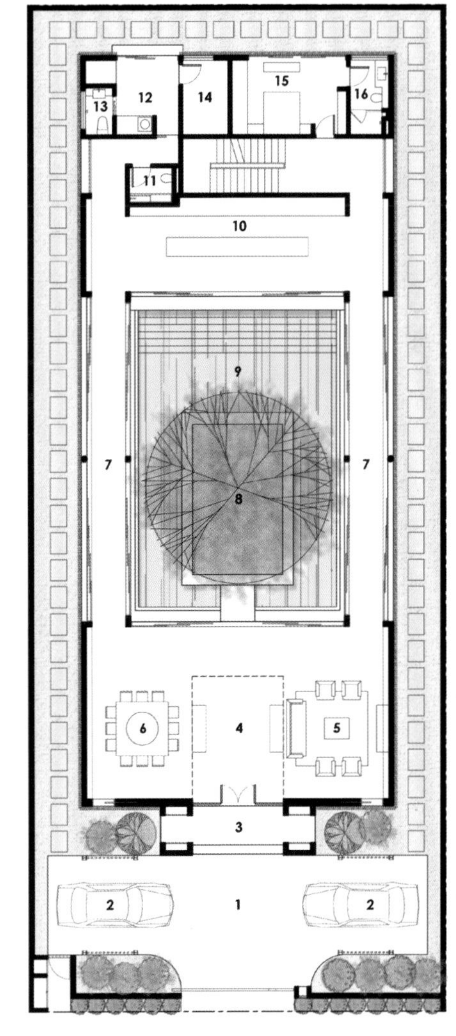

1st Storey Plan　　一层平面图

1. Driveway
2. Carporch
3. Entrance Porch
4. Foyer
5. Living
6. Dining
7. Linkway
8. Courtyard
9. Swimming Pool
10. Open Kitchen
11. Powder Room
12. Laundry
13. Toilet
14. Utility
15. Bedroom
16. Bath

1. 车道
2. 车库
3. 入口
4. 门厅
4. 客厅
6. 餐厅
7. 走廊
8. 庭院
9. 泳池
10. 开放式厨房
11. 化妆间
12. 洗衣房
13. 卫生间
14. 功能房
15. 卧室
16. 浴室

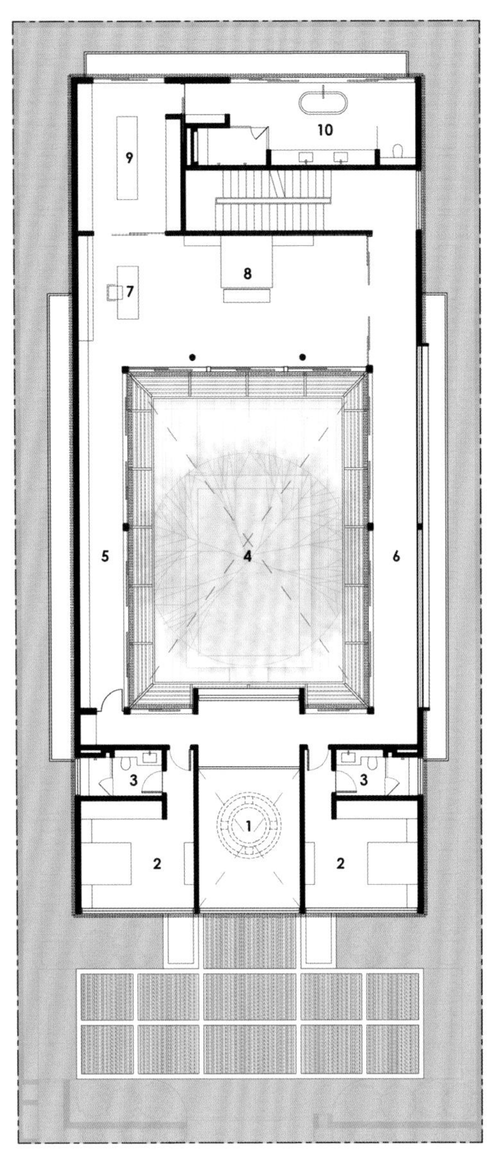

1. Void To Foyer
2. Bedroom
3. Bath
4. Void To Courtyard
5. Library
6. Linkway
7. Study
8. Master Bedroom
9. Walk-In Wardrobe
10. Master Bath

1. 门厅入口
2. 卧室
3. 浴室
4. 庭院入口
5. 图书馆
6. 走廊
7. 书房
8. 主卧
9. 衣帽间
10. 主浴室

2nd Storey Plan　二层平面图

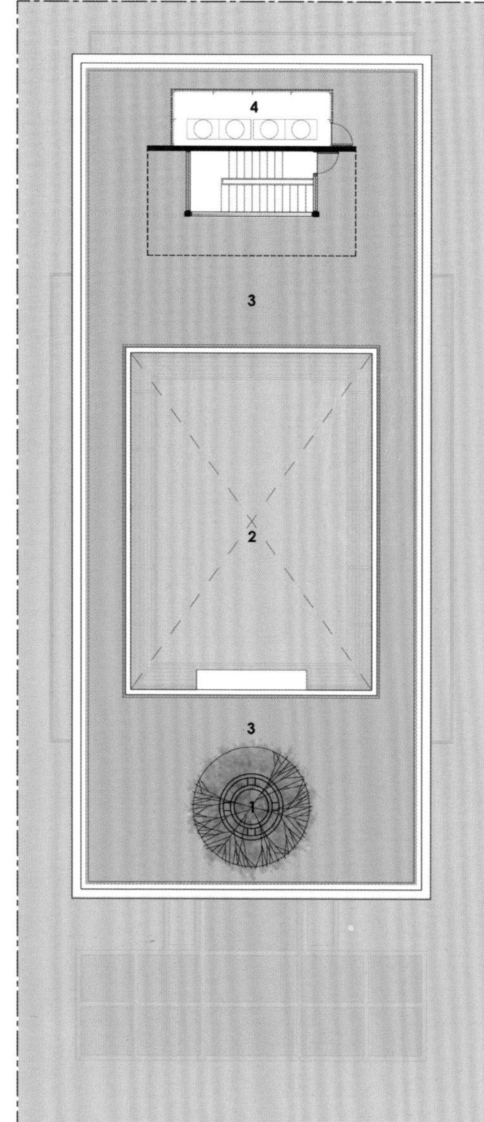

1. Planter
2. Void To Courtyard
3. Timber Deck
4. Air-Con

1. 植物
2. 庭院入口
3. 木质平台
4. 空调机

Roof Plan　屋顶层平面图

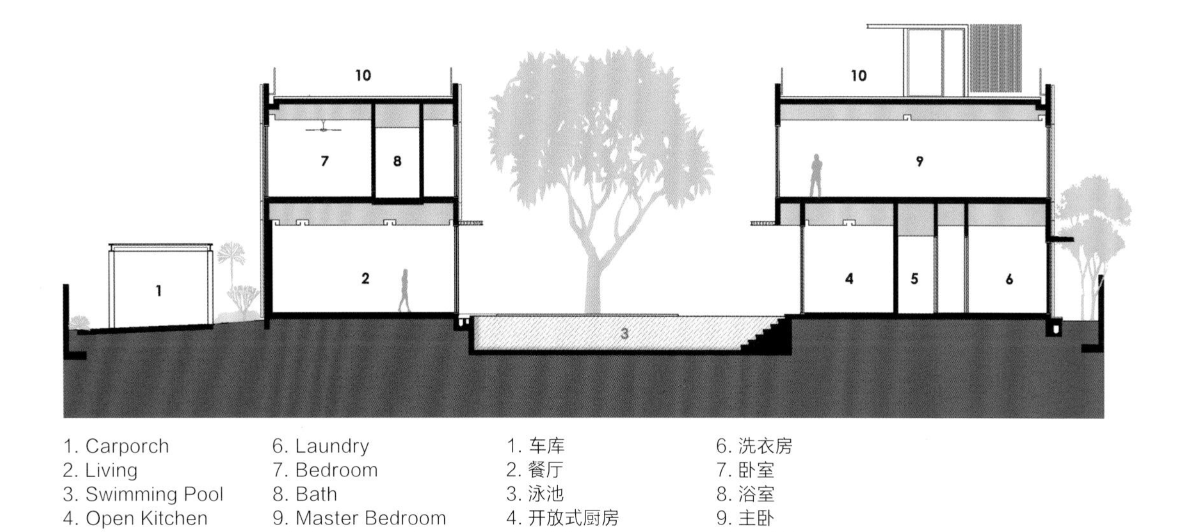

1. Carporch
2. Living
3. Swimming Pool
4. Open Kitchen
5. Powder Room
6. Laundry
7. Bedroom
8. Bath
9. Master Bedroom
10. Timber Deck

1. 车库
2. 餐厅
3. 泳池
4. 开放式厨房
5. 化妆间
6. 洗衣房
7. 卧室
8. 浴室
9. 主卧
10. 木质平台

Section A　剖面图 A

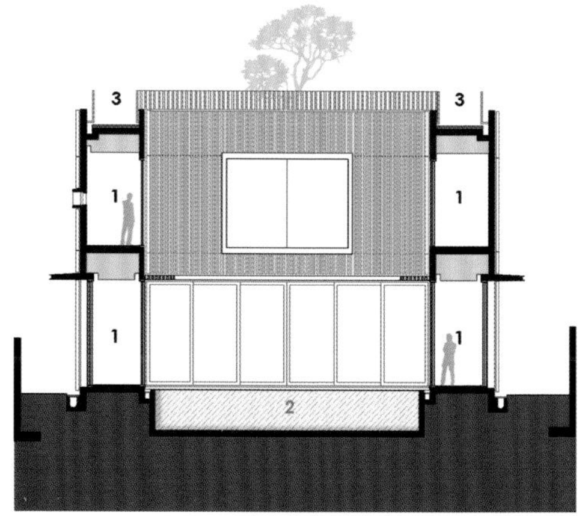

1. Linkway
2. Swimming Pool
3. Timber Deck

1. 走廊
2. 泳池
3. 木质平台

Section B　剖面图 B

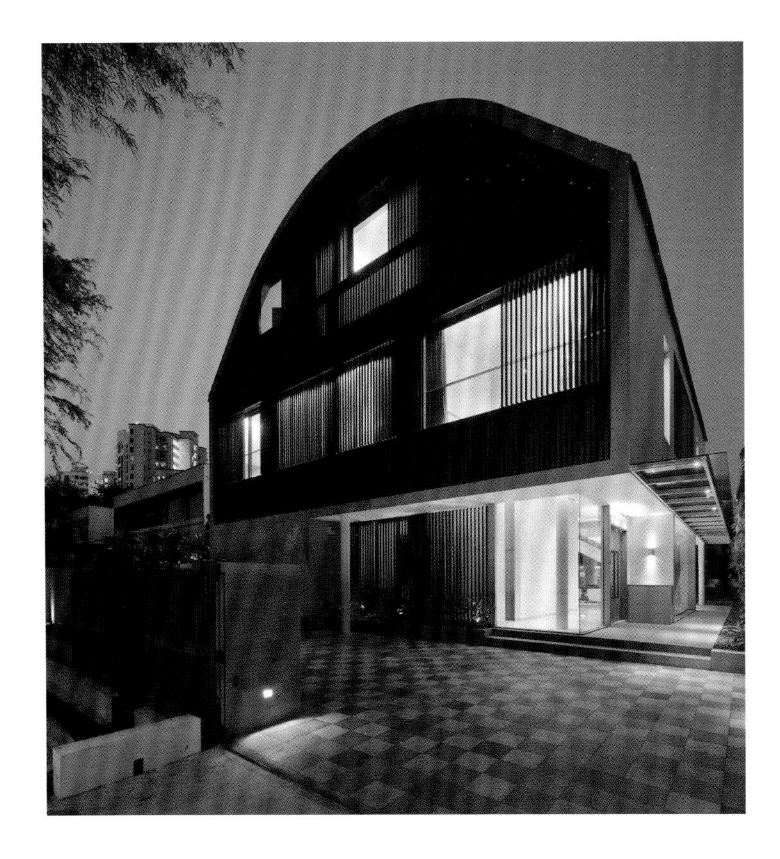

Location: East Coast, Singapore

Architectural Design: Wallflower Architecture + Design

Design Team: Cecil Chee, Robin Tan, Sean Zheng, Shirley Tan & Eileen Kok

Land Area: 553 m^2

Floor Area: 612 m^2

Photography: Jeremy San

项目地点：新加坡东海岸

建筑设计：新加坡桂竹香建筑 + 设计

设计团队： Cecil Chee, Robin Tan, Sean Zheng, Shirley Tan & Eileen Kok

占地面积：553 m^2

建筑面积：612 m^2

摄影：Jeremy San

Keywords 关键词

Tubular Structure 钢管结构

Timber Screens 木质屏风

Visually Expansive 视觉膨胀

Wind Vault House

风拱顶房屋

Overview and Structure

As the brief was substantial, the overall form of the house needed to be pushed to the envelope limits. Naturally, there are also other considerations: the context and proximity of neighbouring homes, the daily sun path and the prevailing winds. Conceptually, the house is a raised reinforced concrete tube whose open ends are oriented in a general north-south direction. On this site, the prevailing breezes also blow in from the south, from the direction of the nearby coast line. In practice, all rooms have walls that side either east or west, and front north and south. The tubular structure resists east west heat gain thanks to the solid mass of the reinforced concrete but encourages passive cooling through the open north south axis. The north and south facades are treated with timber screens and their contribution is multifold.

Architectural Design

Family gatherings are a daily occurrence, and large celebratory functions also feature regularly for the client. The first storey is designed to be visually expansive and uncluttered, and encourages the intermingling of space, whether one moves through the garden, living or dining. The perceived spatial boundary is not architecturally delineated in a traditional sense, but by a soldier-line of narrow polyalthia trees along the boundaries of the site. However, architecturally, the trees provide a level of purpose no man-made wall is able to. It shields the house from neighbouring windows but its narrow footprint still allows sun to reach the grass, while freeing up space on the lawn for play and parties. What may not seem so obvious though is that the tines of polyalthias are literally evaporative fingers, combing the air of some of its heat each time the wind blows. The swimming pool is placed centrally between garden and living both as a focal centre and also as a central evaporative cooling surface.

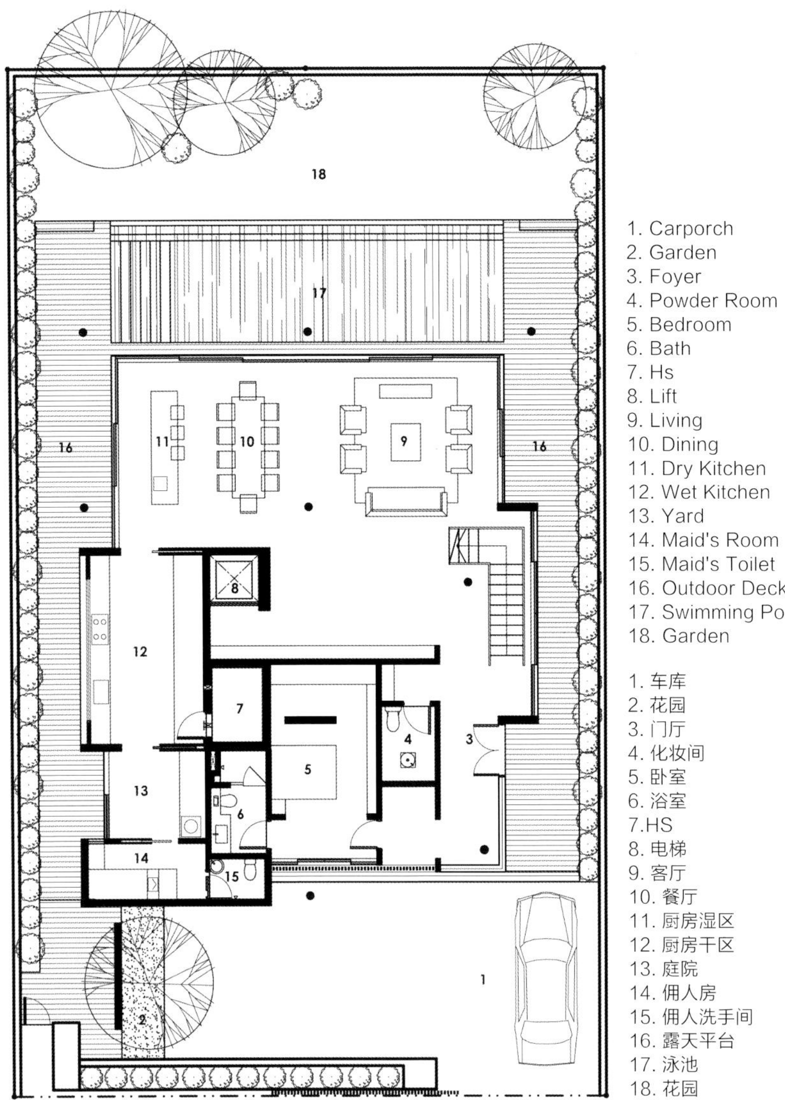

1st Storey Plan　一层平面图

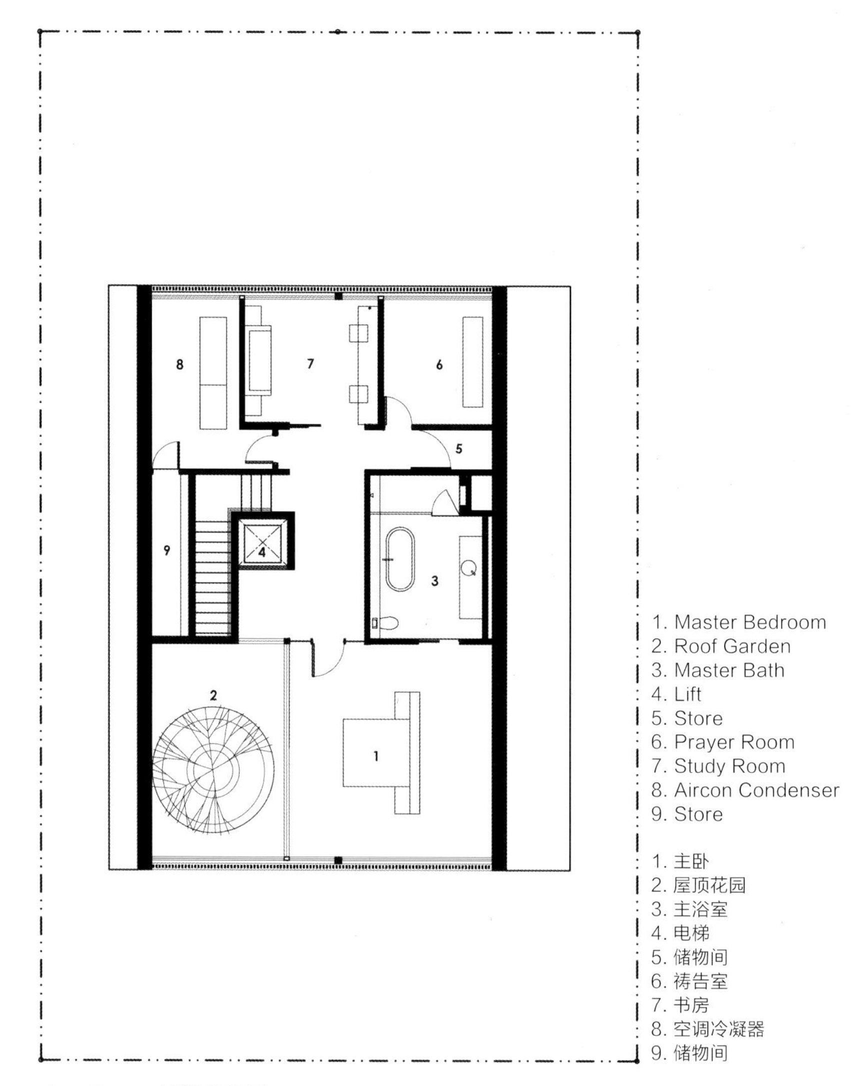

Attic Plan　阁楼层平面图

项目概况与结构

整座房子的建筑形式除了需要满足外部结构最大化这一建筑形式外，还要考虑其他一系列相关的气候环境因素，比如邻居房子的设计形式和布局，每日的太阳路径和盛行风的方向。从建筑结构角度看，这座房子是一个预制钢筋混凝土钢管结构，钢管开口处朝向东南方向，沿岸的海风可以顺着南方吹进屋内。所有的房间几乎都设有朝东或者朝西面的墙体，前庭则是朝北或者朝南的。管状结构可以有效地消除和抵制固体钢筋混凝土在东西立面上产生的热量，同时满足南北轴立面上的被动冷却空调装置。北立面和南立面是由木质屏幕组成的，且安装方式各异。

建筑设计

对于住宅建筑来说，承担诸如家庭聚会、客户晚会等活动都是非常日常化的活动。所以一层大厅在设计上采用了视觉膨胀的效果，从而创造出一种空间交流的变化流动感，整个环境明亮而简洁。感知到的空间边界不是在建筑上传统意义的划定，而是沿着场地边界的一排士兵列队一样的暗罗属树木。从建筑的角度看，原有的树木提供了一种人造墙所不能带来的效果。它阻挡了隔壁窗户，但其狭窄的间隙仍然允许阳光照射到草地上，同时腾出空间可供在草坪上玩耍和聚会。然而似乎不那么明显的是暗罗属树的树枝末梢简直就像无形的手指，每次风吹起来时梳理着一些热气流。游泳池位于花园和生活区两者的中央焦点，同时也是一个集中式蒸发冷却面。

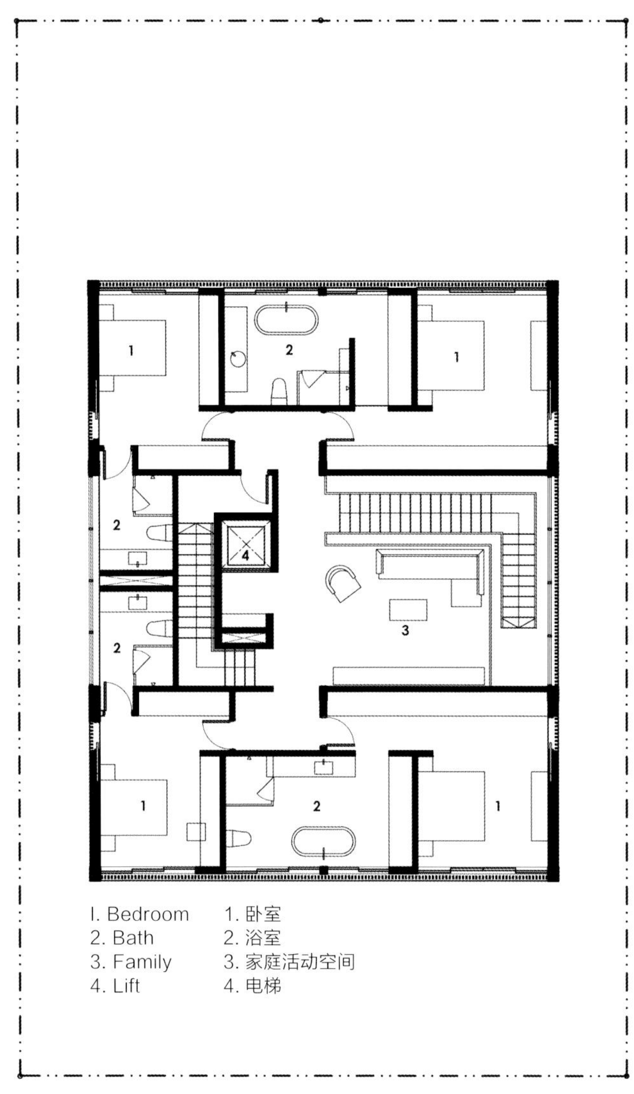

2nd Storey Plan　二层平面图

Location: Jln Angin Laut, Singapore

Architectural Design: Hyla Architects

Main Contractor: 21 Construction Engineering Pte Ltd

Structural Engineer: GNG Consultants Pte Ltd

Land Area: 513.7 m^2

Floor Area: 582.14 m^2

Photography: Derek Swalwell

项目地点：新加坡 Jln Angin Laut 路

建筑设计：新加坡 Hyla 建筑事务所

承包商：21 建筑工程有限公司

结构工程：GNG 咨询有限公司

占地面积：513.7 m^2

建筑面积：582.14 m^2

摄影：Derek Swalwell

Keywords 关键词

Sleek Facade 光滑立面

Patio Design 天井设计

Glass Bridge 玻璃桥

Jln Angin Laut

Jln Angin Laut 路住宅

Overview

Presenting a sleek facade to its neighbors, the finely detailed screens of Jln Angin Laut conceal a house nestled gently into a garden.

Patio Design

Its entrance elevated above the ground, one has to ascend a glass staircase to enter the house. Opening the solid timber front door, one is greeted with a swimming pool and patio surrounded by lush greenery, amply shaded overhead but admitting light and air from the sides. This space is a paradigm of living comfortably in the tropics.

Design Features

A glass bridge spanning lightly across the pool leads into the living room. This bridge extends the threshold of the house, prolonging the act of entering and highlighting the importance of this space to the overall design of the house. The rest of the house takes its cues from this scene, the main living spaces being punctuated with light, greenery and timber accents. Together with the skillful manipulation of solids and voids, the overall effect achieved is that the architecture seems integrated harmoniously with nature.

19

Basement Plan　地下层平面图

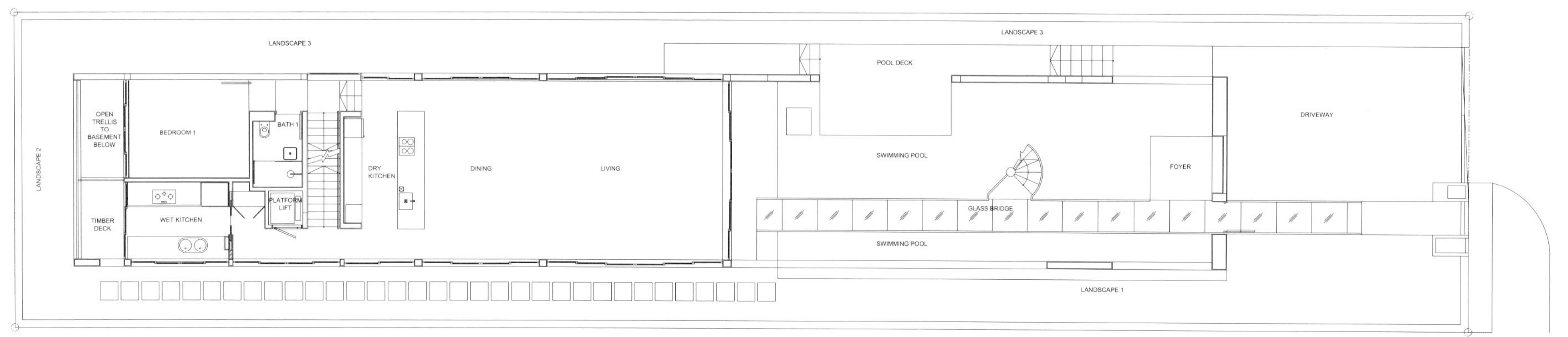

First Storey Plan　一层平面图

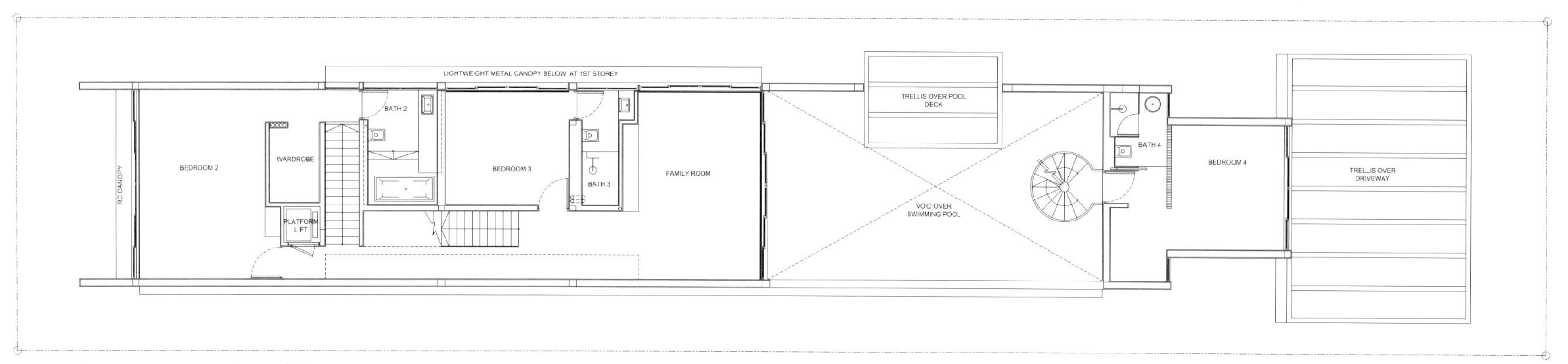

Second Storey Plan　二层平面图

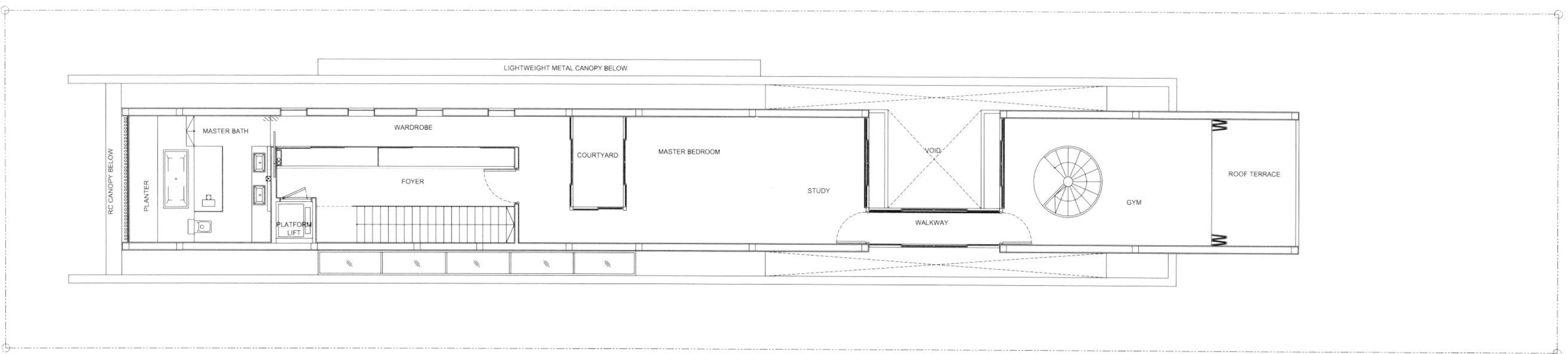

Third Storey Plan　三层平面图

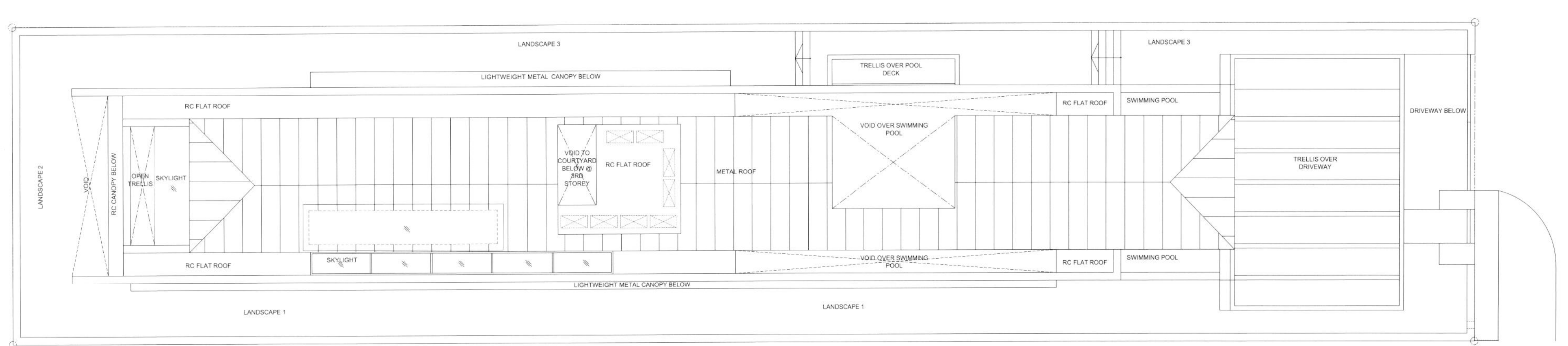

Roof Plan　屋顶层平面图

项目概况

这个外表光滑的建筑物就是Jln Angin Laut 住宅。在它的内部其实还隐藏着一个与世隔绝、静谧的小花园。

天井设计

住宅的入口处高出地面，需要走上一段玻璃台阶才能进入住宅内部，打开实木大门，迎接人们的是一个天井，下面是游泳池，而四周则是茂盛的热带植物。天井阻挡了炎热的阳光，同时也保证了空气流通和自然采光。这种天井空间设计可以说是热带住宅设计中较好地满足了舒适度的范本。

设计亮点

最有特点的要数泳池上方的玻璃桥了，必须从水面上走过才能到达真正的生活居住空间。水域将住宅与外部完全隔离开来，而透明的玻璃桥又延长、加强了这种隔绝感，为私人空间的营造做足了功夫。入口部分的设计可以说是整个住宅设计中最大的亮点，而且其他部分的设计都是围绕着入口的天井展开的。其余的居住空间不时地被光影或者绿色植物“打断”，营造出长条形住宅的节奏感，避免了空间的单调乏味。其中还运用巧妙的设计，形成虚实的对比，整体上将建筑很好地融入自然环境中。

Semi-detached Houses
半独立式别墅

Unique Shape　Elegant Courtyard　Beautiful Environment

造型独特　庭院优雅　环境美观

Location: Lasia Avenue, Singapore

Architectural Design: Aamer Architecs

Civil & Structural Engineer: J S Tan & Associates

Quantity Surveyor: BKG Consultants Pte Ltd

Contractor: KLT Builders Pte Ltd

Landscape Design: Tropical Environment Pte Ltd

Land Area: 387 m^2

Floor Area: 393 m^2

Photography: Amir Sultan

项目地点：新加坡 Lasia 大道

建筑设计：Aamer Architects

土建工程：J S Tan & Associates

工料测量：BKG 顾问有限公司

承包商：KLT 建筑公司

景观设计：Tropical Environment 有限公司

占地面积：387 m^2

建筑面积：393 m^2

摄影：Amir Sultan

Keywords 关键词

Simple and Effective 简洁高效

Cylindrical Volume 圆柱形空间

Transparent and Bright 光亮通透

Lasia
Lasia 住宅

The plot of this semi-detached house is very small, the design strategy adopted has to be simple and effective. The main design strategy was to pull the front part of the house away from the party wall and express it as a chamfered cylindrical volume covered with vertical timber louvers, which the architects nicknamed "the lipstick". Visually, this gave the house an identity distinct from its neighbour. The space between "the lipstick" and the party wall shared with the neighbour became a circulation zone, housing the car porch and the staircase. To emphasize the separation between "the lipstick" and the party wall, this zone was designed to be transparent and light. A steel, glass and timber structure was thus used which created intriguing shadow and light patterns.

这栋半独立式住宅的地块非常小，因此设计方案必须简洁高效。这个方案的重点是将房子的前面部分远离界墙，加上垂直的木制百叶窗，使其成为一个有棱有角的圆柱形空间。设计师们称它为“口红房子”。在外形上，这使得该住宅与周边截然不同。“口红房子”与界墙之间的空间成了一个流通带，提供了车辆通道和楼梯的空间。为了强调“口红房子”与界墙的分界线，这个流通带设计得光亮透明，拥有钢材、玻璃与木材相结合的构造，也创造出有趣的投影和光图像。

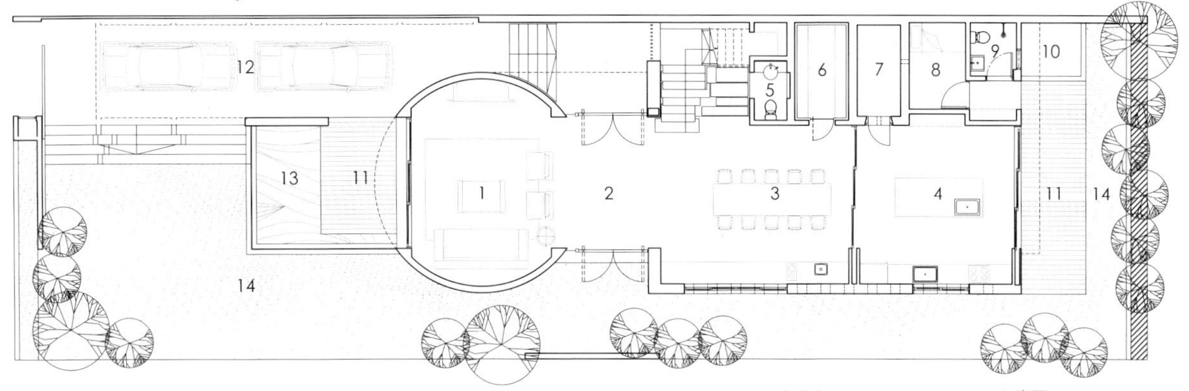

1st Storey Plan 一层平面图

1. living room	1. 客厅
2. foyer	2. 门厅
3. dining/kitchen	3. 餐厅 / 厨房
4. wet kitchen	4. 厨房湿区
5. powder room	5. 化妆间
6. wine cellar	6. 酒窖
7. household shelter	7. 遮蔽区
8. maid's room	8. 佣人房
9. maid's bath	9. 佣人浴室
10. washing area	10. 洗漱区
11. terrace	11. 露台
12. carporch	12. 车库
13. pond	13. 池塘
14. garden	14. 花园

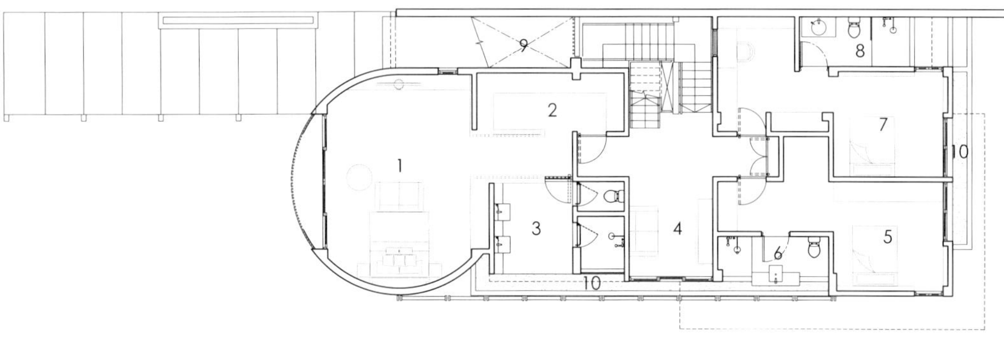

2nd Storey Plan 二层平面图

1.Master Bedroom	1. 主卧
2. Walk-In Wardrobe	2. 衣帽间
3. Master Bath	3. 主卧浴室
4. Familyroom	4. 家庭室
5. Bedroom 1	5. 卧室 1
6. Bathroom 1	6. 浴室 1
7. Bedroom 2	7. 卧室 2
8. Bathroom 2	8. 浴室 2
9. Void Over Carporch	9. 车库入口
10. Garden	10. 花园

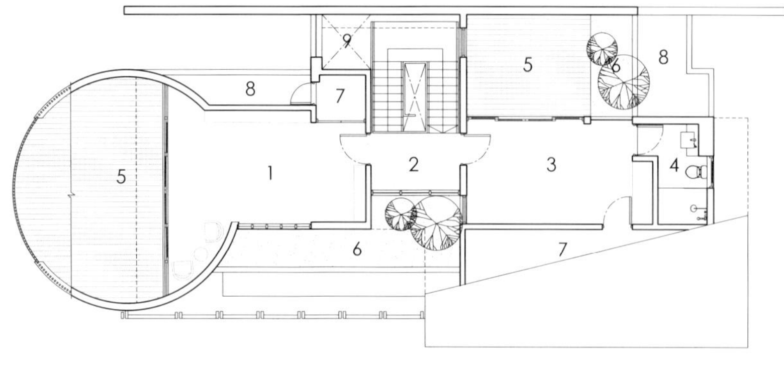

Attic Storey Plan 阁楼层平面图

1. Gym	1. 健身房
2. Foyer	2. 门厅
3. Study	3. 书房
4. Bath3	4. 浴室 3
5. Terrace	5. 露台
6. Roof Garden	6. 屋顶花园
7. Store	7. 储物间
8. A/C Ledge	8. 空调架

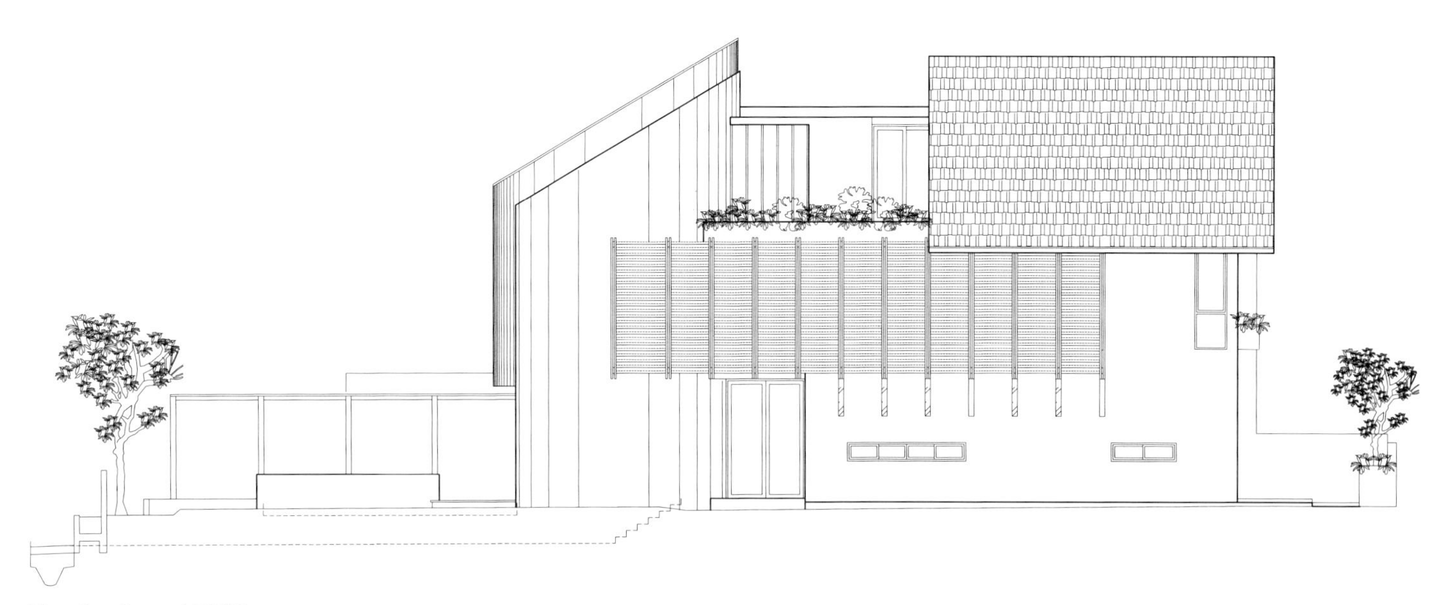

Elevation 2　　立面图 2

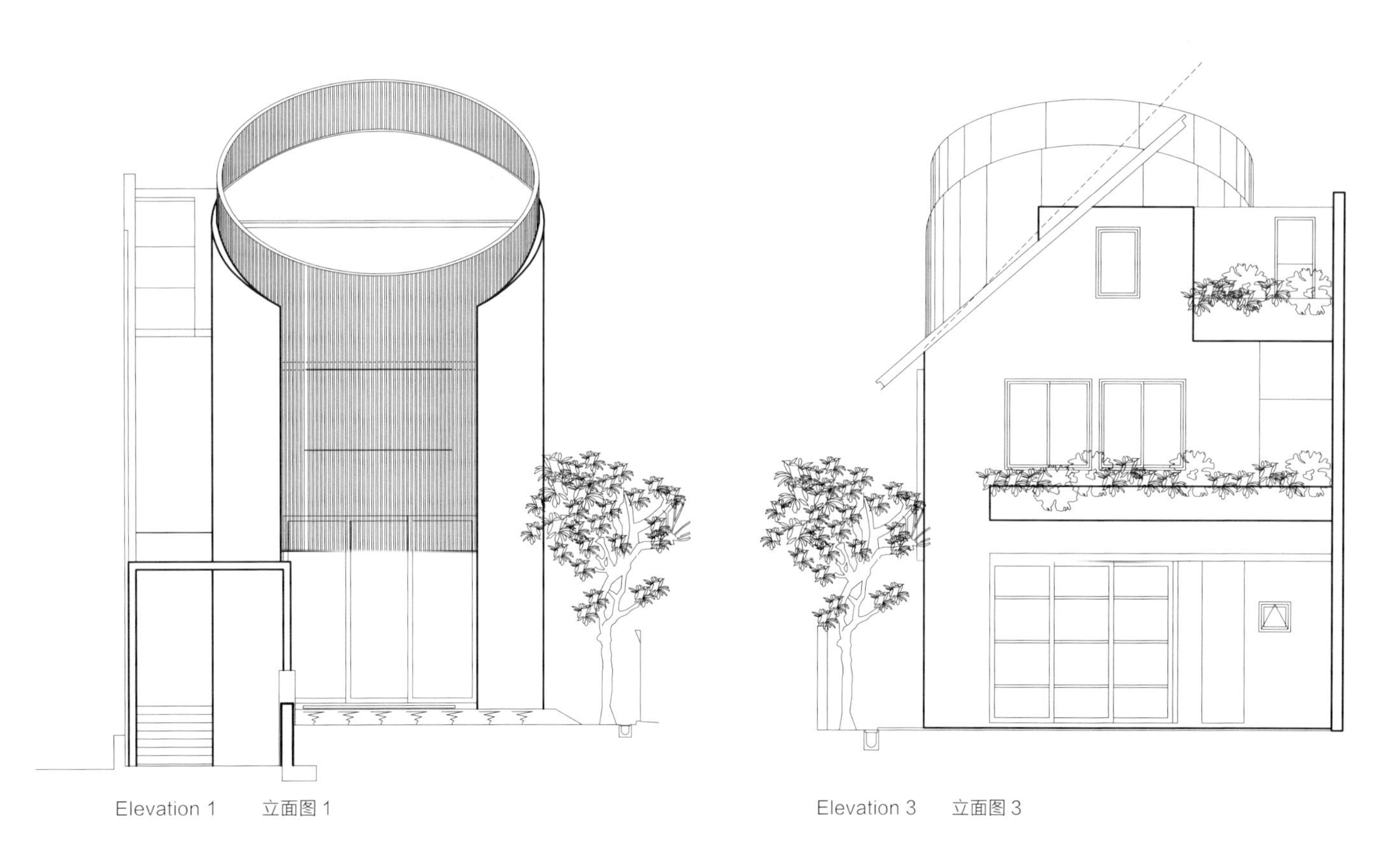

Elevation 1　　立面图 1

Elevation 3　　立面图 3

Location: Serangoon, Singapore

Architectural Design: Wallflower Architecture + Design

Design Team: Robin Tan, Cecil Chee, Sean Zheng, Shirley Tan, Eileen Kok & Wichupon

Land Area: 367 m^2

Floor Area: 436 m^2

Photography: Marc Tey

项目地点： 新加坡实龙岗

建筑设计：新加坡桂竹香建筑 + 设计

设计团队：Robin Tan, Cecil Chee, Sean Zheng, Shirley Tan, Eileen Kok & Wichupon

占地面积：367 m^2

建筑面积：436 m^2

摄影：Marc Tey

Keywords 关键词

Linear Space 线性空间

Open View 开阔视野

Natural Ventilation 自然通风

Sunny Side House

向阳住宅

Overview

The site would not appeal to most local homebuyer as it immediately ticks several negative boxes for what are deemed liabilities in a residential semi-detached plot. It is long and narrow, with both the long side and front facing the western afternoon sun. The plot lies a metre below a public road that bounds the front and the 'sunny' side.

Design Concept

The clients, a family of five, wanted a home that revolved around familial living and bonding. The implication of this ideal is that spaces need not be arranged or defined too rigidly for formal or cultural hierarchies. There are advantages to be had from a narrow plot as it naturally restricts the depth of rooms. The resulting spaces receive more natural light and are better ventilated due to the shallower proportion.

Linear Space Design

The 1st storey is conceived to be a contiguous linear space, where living, dining, kitchen functions are serially arranged, but have little in the way of physical demarcation. Back of house spaces that need enclosing walls are aligned against the inner party wall and do not intrude into the informal living/dining/kitchen. In an Asian context, it is perceived as undesirable when a home is set lower than the surrounding public areas; mainly due to connotations of a lower status. Architecturally however, the lowered topography and resulting upturn along the lengthy boundary edge helps to delineate and suggest that the living space extends into the green.

THREE GENENG

Bedroom Design

The second storey is the bedroom box, the rooms arranged in a row along the outer edge. Each room has generous private views out. The substantial heat gain from an afternoon of harsh tropical sun is mediated by the extensive use of timber screens beyond the openings. The strategy is not to fully block out but filter harsh light and heat. It vitally allows natural light and ventilation to still pass through. It enables the building to breath, airflow being essential to comfort within the tropics. The pivoted screens also double up to control visual privacy, and can be manually angled to adjust for individual preferences. The visually solid form of the 2nd storey is given texture and interest by the timber fins, alleviating what would be an oppressive block facing the public street.

Top Floor Design

The top floor is deliberately 'lightened', surrounded by full glass fenestration supporting a deep, overhanging roof. It reduces the visual mass from a compositional point of view, but also allows unobstructed enjoyment to distant views. A section of the top floor is once again a contiguous space for recreation. The entertainment space is not just for movies but more importantly for football league matches faithfully followed by friends and family, a regular event in this household. A foosball table and an open pantry support the pre and post-game excitement and a convenient guest room is available in case some decide to stay on.

Corridor and Stairs Design

Key to linking and encouraging full use of the house at all levels are the inboard corridors and stairs. The aesthetics are intentionally kept minimal and uncluttered. Numerous indirect skylights filter and bathe this multilevel space in light. The open thread stairs, slender handrails, stringers and frameless glass panels facilitate visual connections from the 1st to the top storey, making the perceptual space and volume larger, always inviting one to explore a different level.

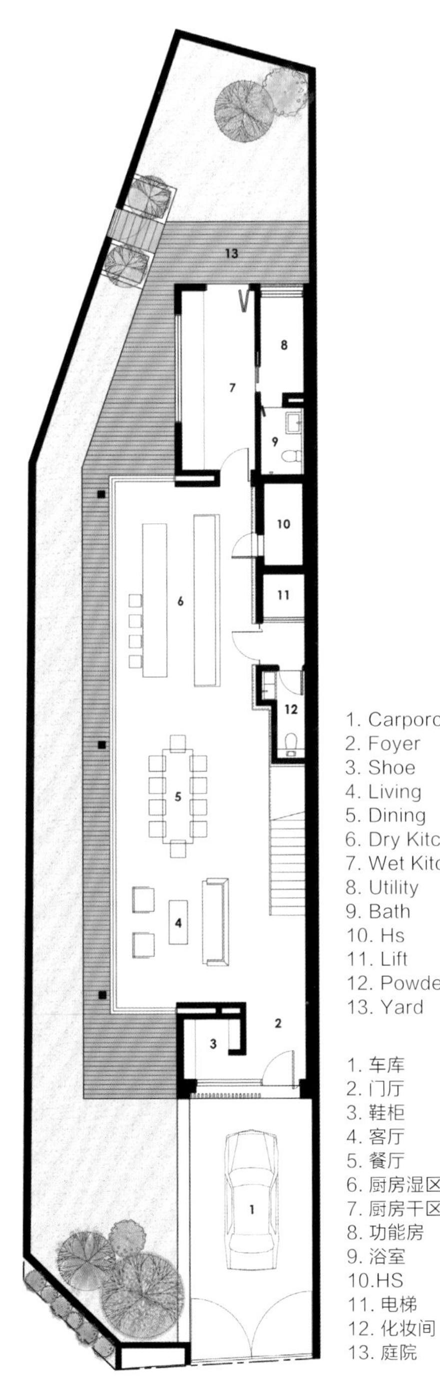

1. Carporch
2. Foyer
3. Shoe
4. Living
5. Dining
6. Dry Kitchen
7. Wet Kitchen
8. Utility
9. Bath
10. Hs
11. Lift
12. Powder Room
13. Yard

1. 车库
2. 门厅
3. 鞋柜
4. 客厅
5. 餐厅
6. 厨房湿区
7. 厨房干区
8. 功能房
9. 浴室
10.HS
11. 电梯
12. 化妆间
13. 庭院

1st Storey Plan　一层平面图

1.Carporch Roof
2. Master Bedroom
3. Master Bathroom
4. Bedroom
5. Bath
6. Lift
7. Store

1. 车库屋顶
2. 主卧
3. 主浴室
4. 卧室
5. 浴室
6. 电梯
7. 储物间

2nd Storey Plan　二层平面图

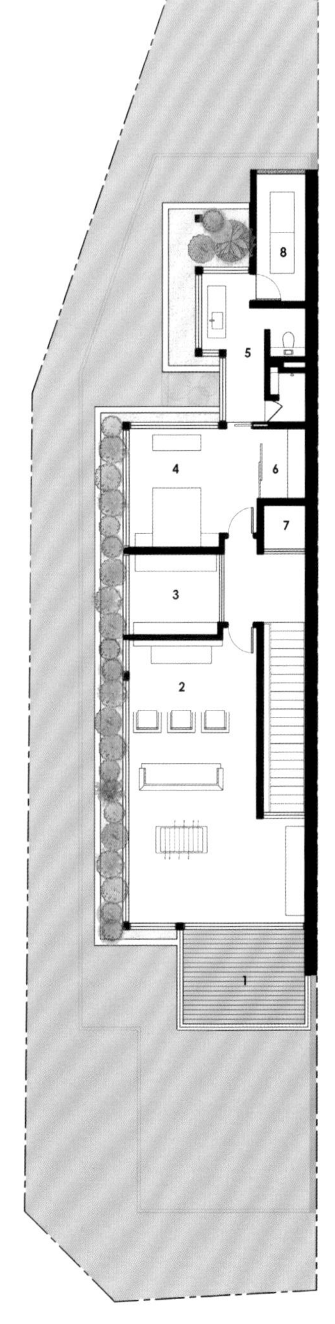

1. Roof Terrace
2. Entertainment
3. Pantry
4. Bedroom
5. Bath
6. Wardrobe
7. Lift
8. Aircon

1. 屋顶露台
2. 娱乐间
3. 食品储藏室
4. 卧室
5. 浴室
6. 衣橱
7. 电梯
8. 空调

3rd Storey Plan　三层平面图

项目概况

该场所将不会吸引很多当地购房者，因为其在半独立式的住宅中被认为是种累赘。房屋长而窄，长边和正面都正对着西部午后阳光。该地块距离公共道路小于 1 m，是正面和向阳面的边界。

设计理念

业主为五口之家，想要一个围绕家庭生活和相互关联的家。这种理想的含义就是，空间不需要局限于太过僵硬的形式或者文化层次。狭长的场地自然而然形成了房间的大小,由此产生的空间能接受更多自然光线和更好的通风。

线性空间设计

一层是一个连续的线性空间，起居、餐厅、厨房按照顺序排列，但并没有物理划分。房屋的后面要有围墙，与房内的界墙对齐，并不影响日常的起居、餐厅和厨房。从亚洲传统来看，一座家园处于比周围环境更低的位置，被看作是不可取的，这主要归咎于较低位置的内涵。然而从建筑方面来看，较低的地形和长长的边界向上延伸，有助于勾画并凸显出向绿色延伸的生活空间。

卧室设计

二层是卧室，房间沿着外缘排成一条。每个房间都享有开阔的私人视野。午后热带严酷的阳光带来的巨大热量通过开口外的树林屏障的广泛使用得以调节。其策略不是完全屏蔽而是过滤强光强热,容许自然光线和自然风穿过，使整栋建筑得以呼吸，热带气流是必不可少的慰藉。旋转屏幕进一步控制视觉隐私，并可以依据个人喜好手动调整角度。第二层的视觉固体形式带有质感，木材散热翅片十分有趣，减少了房屋面临公众街区的压迫感。

顶层设计

顶层刻意“增亮”，用整块玻璃作窗，支撑起向外延伸的屋顶。从构图的角度来看，减少了视觉效果，但是可以畅通无阻地享受远处的风景。顶层的一部分再一次延续了休闲的空间。娱乐空间不仅是为了看电影，更重要的是忠实球迷的朋友家人日常可以观看足球联赛。一张桌上足球桌和一个开放式餐柜足以满足赛前赛后的所有需求，还有一间备用的客房方便有人决定留宿。

走廊与楼梯设计

连接和充分利用房屋每个层次的关键是室内的走廊和楼梯。出于美学考虑，有意保持极简艺术和整洁。充足的天光间接过滤下来，这个多层次的空间都沐浴在阳光下。开放的线形楼梯、细长的扶手、纵梁以及无框玻璃嵌板造成了一层到顶层的视觉上的联系，扩大感知到的空间和体积，邀人进一步探索一个不同的层面。

Location: Telok Kurau, Singapore

Architectural Design: A D Lab pte ltd

Design Team: Warren Liu, Yaw Lin

Floor Area: 547m^2

Photography: Edward Hendricks, CI&A

项目地点：新加坡直落古楼

建筑设计：A D Lab 设计有限公司

设计团队：Warren Liu, Yaw Lin

建筑面积：547 m^2

摄影：Edward Hendricks, CI&A

Keywords 关键词

Semi-detached 半独立式

Brick Wall 砖墙

Natural Ventilation 自然通风

16A Lorong M Telok Kurau

直落古楼 M 巷 16A 别墅

Overview

A quiet and unassuming gesture to the street with a respect of the humble scale of the adjacent series of semi-detached houses along Lorong M Telok Kurau describes the entrance to this reconstruction project by Dlab in Singapore. With only a second glance, one might notice the subtle moiré pattern beginning to emerge on the brick work of the front facade that would introduce the main design intention of the house and define its relationship with the surrounding land and context. In this project, the architects were careful not to waste what was already on the site. Architects retained the structure of the existing semi-detached building.

Architectural Design

As is often the case with semi-detached houses in Singapore, the long open side of the house faces the long side of the neighbour's property, creating an "overlooking" problem. Here, the archietcts' resolution to this problem was to look back at a basic element of architecture— the brick. The entire side of the existing structure of the Telok Kurau building was wrapped by the designers in a new organic skin made of brick that stretches open in areas to let light and ventilation through and closes where privacy is required. The resulting language is a timeless rethinking of traditional materials. The site is a very deep and relatively narrow plot of land that faces on its longest side of an existing towering 4 storey building that is built up to the maximum setback. Because of the extended depth of the site, the architects chose to perform some minor surgery on the existing structure by breaking it open in its center to allow an internal atrium to unify the building vertically. This atrium also introduces light and long diagonal views up through from the Living and Dining Rooms on the first floor to the master bedroom and family room on the 2nd floor and further up to the master study and music room in the attic. Along the entire ground floor, a band of full heights glass windows links the series of spaces together with a constant relationship to the narrow strip of garden at the side. Since the garden on the ground floor could be made to be quite private with the clever use of planting and fencing, the rooms open up as much as possible to the external to allow the maximum amount of light to enter the long spaces.

From the second floor onward, the effect of the elegant brick patterns can be experienced. The large expanses of patterned brick walls that open in gradual organic sequences are separated from the external spaces with sheets of sliding glass that can be moved aside. Beautiful and hypnotic reflected patterns of light from the brick screen fall lightly on the floors and surrounding walls, giving the spaces an almost cathedral-like quality and an ambiguous semi outdoor character.

With this project at Telok Kurau, Architects managed through simple, inexpensive and effective design to reduce waste, to work within an economy of means, as well as to create beautiful and poetic space that is beneficial to the inhabitant as well as to the neighbours and the surrounding natural environment.

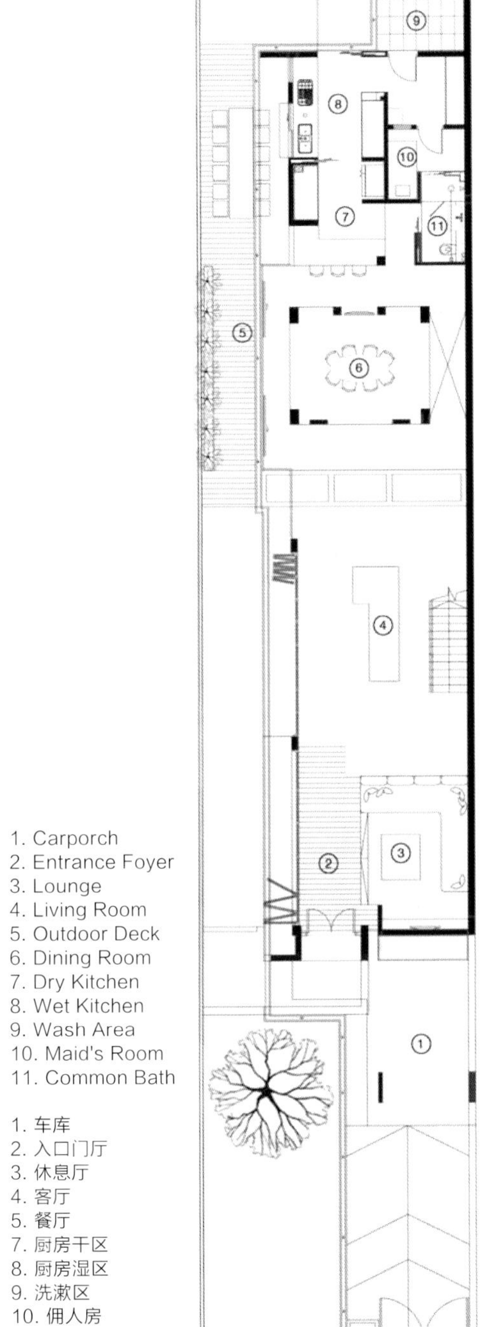

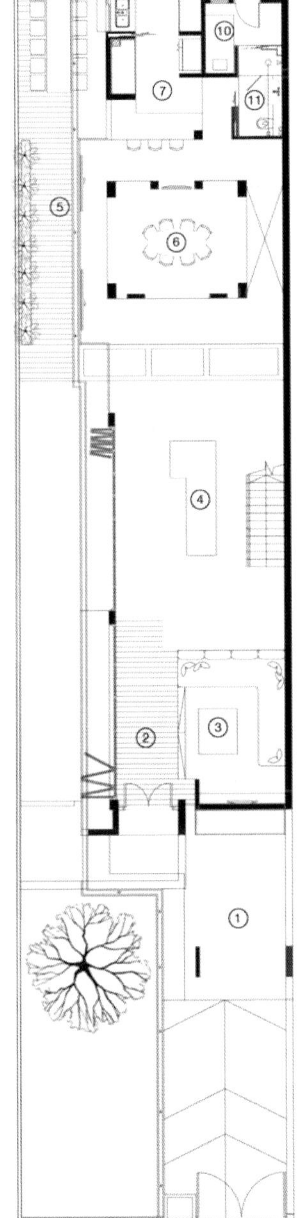

1. Carporch
2. Entrance Foyer
3. Lounge
4. Living Room
5. Outdoor Deck
6. Dining Room
7. Dry Kitchen
8. Wet Kitchen
9. Wash Area
10. Maid's Room
11. Common Bath

1. 车库
2. 入口门厅
3. 休息厅
4. 客厅
5. 餐厅
7. 厨房干区
8. 厨房湿区
9. 洗漱区
10. 佣人房
11. 公共浴室

Storey 1 Plan　一层平面图

12. Outdoor Deck
13. Master Bath
14. Master Bedroom
15. Family Room
16. Common Bath Two
17. Bedroom One
18. Bedroom Two

12. 室外平台
13. 主卧浴室
14. 主卧
15. 家庭间
16. 公共浴室 2
17. 卧室 1
18. 卧室 2

Storey 2 Plan　二层平面图

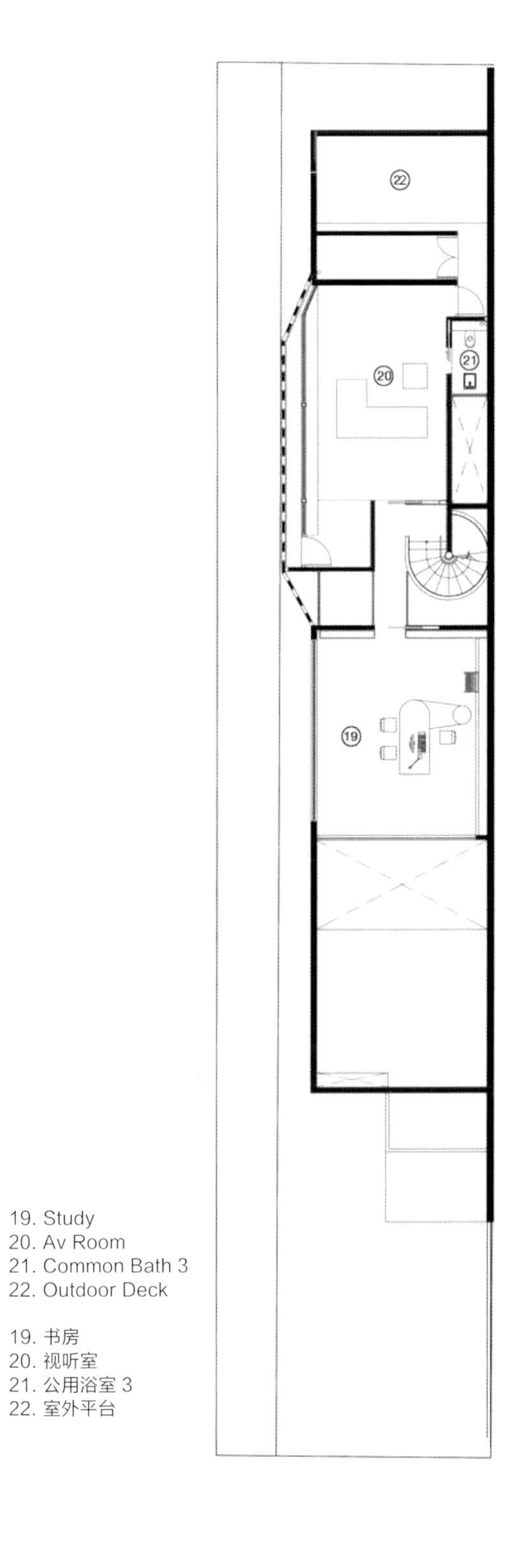

19. Study
20. Av Room
21. Common Bath 3
22. Outdoor Deck

19. 书房
20. 视听室
21. 公用浴室 3
22. 室外平台

Attic Plan　阁楼层平面图

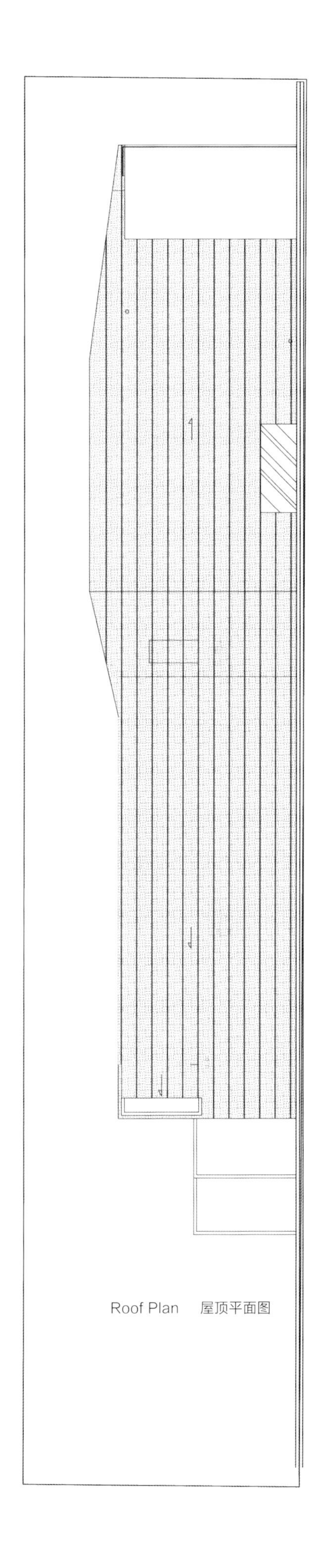

Roof Plan　屋顶平面图

项目概况

由AD Lab公司改建的这栋别墅，与相邻的一系列沿直落古楼路M巷的半独立式别墅以安静、不张扬的姿态面向街道。只需再看一眼，就会注意到微妙的摩尔纹图案浮现在前立面的砖头上，展现出别墅主要的设计意向，设定了其与周围环境的关系。在这个项目上，建筑师们十分注意不浪费原址已有的材料，保留了原来的半独立式建筑结构。

建筑设计

新加坡半独立式别墅的普遍问题是相邻别墅都是长长的，造成了“远眺”的障碍。建筑师们为了解决这一问题，在建筑最基本的元素——砖上寻求答案。直落古楼这栋别墅现存结构的整个侧边都被设计师围绕在一个砖块组成的新的表层，并向外延伸，可以采光和通风，同时很好地保护隐私。建筑师在建筑过程中始终关注传统材料的运用。别墅位于一块深长狭窄的土地上，其最长侧的一面所面临的是一栋高达四层的建筑物，这是建筑过程中最大的障碍。考虑到别墅外延的深度，建筑师决定对现有结构做些微调整，将中间打通，使室内的中庭与建筑垂直统一。这样的中庭还可以引入光线，形成从一层的起居室和餐厅到二楼卧室的对角线视野，更远一点可以看到阁楼的书房和音乐室。整个一楼四周，用全高度的玻璃连接起一系列的空间以及侧边的狭窄地带的花园。一楼的花园巧妙地栽植围绕成栅栏，房间尽可能多地向外开放，既可最大限度保护安全隐私又能采光。

从二楼上去，可以体验讲究的砌砖式码放带来的效果。有图案的砖墙逐渐有序地大面积铺开，室内空间用滑动玻璃门隔开。从砖墙反射而来的漂亮又柔和的光线轻轻地落在地板和四周的墙上，营造出一种几近大教堂一般的氛围和模糊的半户外特征。

这座别墅成功地通过简单价廉但十分有效的设计减少了浪费，经济的工作方式同时创造出美丽如诗般的空间，对居民、邻居以及周围的自然环境十分有益。

Location: Bukit Batok, Singapore

Architectural Design: A D Lab Pte. Ltd.

Design Team: Warren Liu Yaw Lin, Najeeb Rahmat, Dennis Ng, Usha Bragenshyam

Civil Structural Engineer: Aston Consulting Engineers

Contractor: JK Integrated Pte Ltd

项目地点：新加坡武吉巴督镇

建筑设计：AD Lab 私人有限公司

设计团队：Warren Liu Yaw Lin, Najeeb Rahmat, Dennis Ng, Usha Bragenshyam

土建工程：埃斯顿咨询工程师

承包商：JK 综合私人有限公司

Keywords 关键词

The 5th Elevation 第五立面

Twisting 弯曲折叠

Semi-detached 半独立式

8 Lorong Kemunchup

Kemunchup 巷 8 号别墅

Overview

With this reconstruction project on Lorong Kemunchup, the designers were faced with an extremely challenging site with very exciting possibilities. An existing conservative semi-detached house sat inconspicuously at the street frontage of the site. Lurking behind this house, however, was a 100 m long narrow strip of land that belongs to the property and tapers gradually toward the back in a boomerang shape. Apart from being extremely long and narrow, this strip of land, which reaches around the base of a surrounding hill in a heavily forested area, slopes approximately 6~7 meters from the back to the front of the strip. Because this long strip was thought to be an unbuildable area, much of the site was left unoccupied and neglected. Out of the 10,000.00 ft^2(929.03m^2) of the site, the existing semi-detached house only covered approximately 30% of its area. Architects saw that this neglected strip of land, despite its treacherous terrain, was the most beautiful portion of the site and challenged themselves to build upon it.

Architectural Design

As a way of reducing waste on the project and of saving cost, the designers decided to retain the entire 2 storey plus attic semi-detached house with minimal refinishing work to be done. the designers concentrated the efforts of the project at the back of the site. Firstly, an existing single storey extension at the back of the semi-detached house was demolished and the ground beams and slab were re-used for the new single storey living room. A second terracing two storey structure was added at the back of the site and separated from the Living Room with a swimming pool and covered walkway and terraces that wrap around it. Due to the fact that the structure is terracing to follow the sloping of the ground form, several roof terraces were created for the inhabitants to enjoy the fantastic surrounding scenery. The new structure was kept low and ground-hugging in order to keep it light and minimize the foundation work. In another effort to minimize material, the designers used the 5th elevation of the house, the roof form, as the main facade of the extension. the designers bent and folded this form around the top and the sides of the house. This roof was conceived as an evolution of the traditional sloping gable tropical roof and retained the idea of the visual and functional importance of the roof in the tropics. Aesthetically, this undulating roof form also echoes the language of the surrounding hills and organically wraps and manipulates itself around the sloping, twisting site.

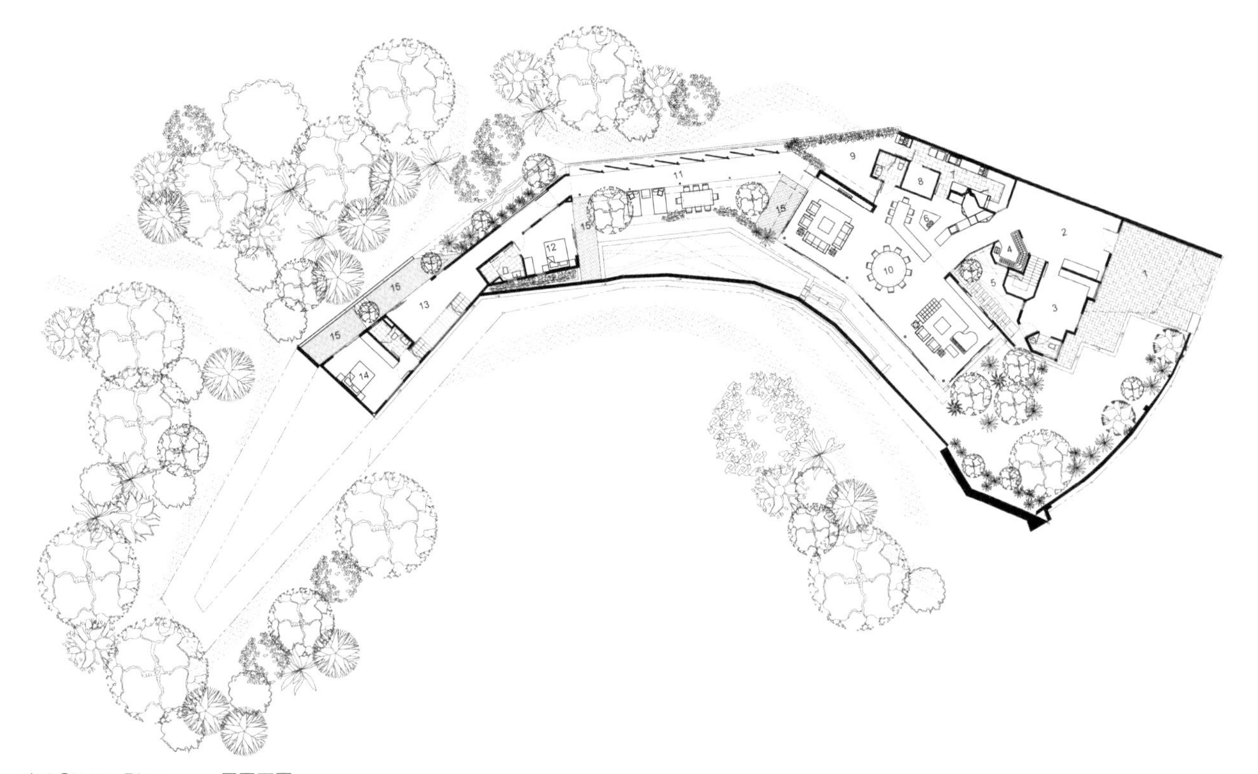

1. Carporch
2. Entrance Hall
3. Study Room
4. Reading Niche
5. Reflecting Pool
6. Dry Kitchen
7. Wet Kitchen
8. Utility Room
9. Yard
10. Family, Living & Dining Room
11. External Linkway
12. New Bedroom 5
13. New Family Room 2
14. New Bedroom 4
15. External Deck

1. 车库
2. 入口大厅
3. 书房
4. 阅读区
5. 镜面池塘
6. 厨房干区
7. 厨房湿区
8. 功能用房
9. 院子
10. 家庭室、客厅、餐厅
11. 外部走廊
12. 新卧室 5
13. 新家庭室 2
14. 新卧室 4
15. 外部平台

1st Storey Plan 一层平面图

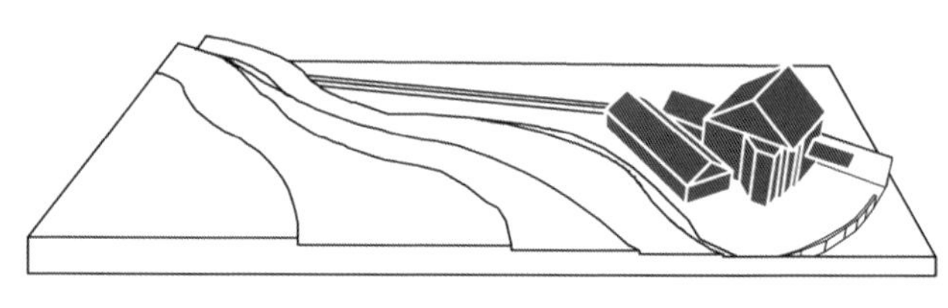

Existing Semi-Detached House
现状双拼住宅

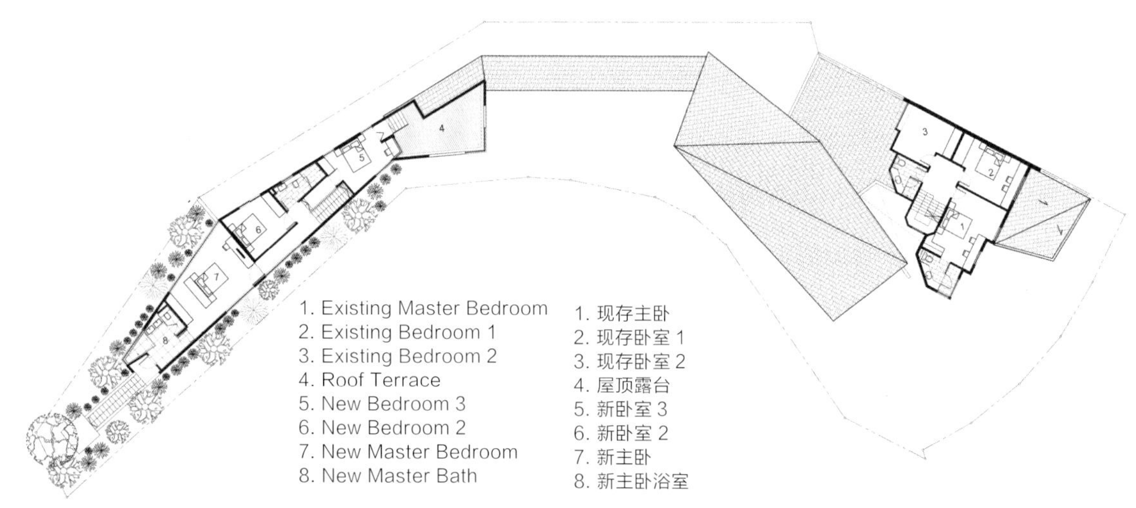

2nd Storey Plan 二层平面图

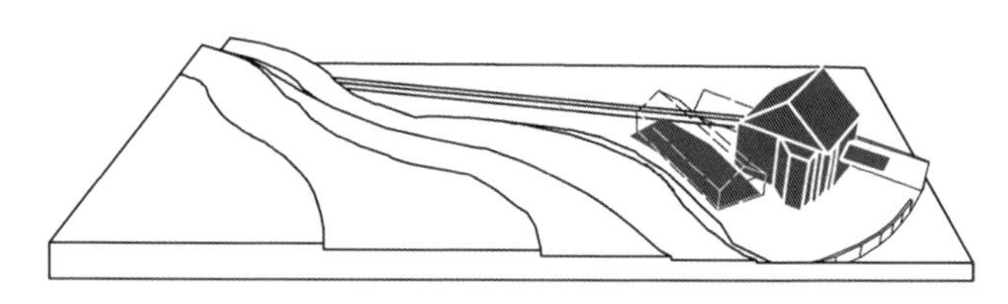

Removal Of Existing Block
现状体块移动

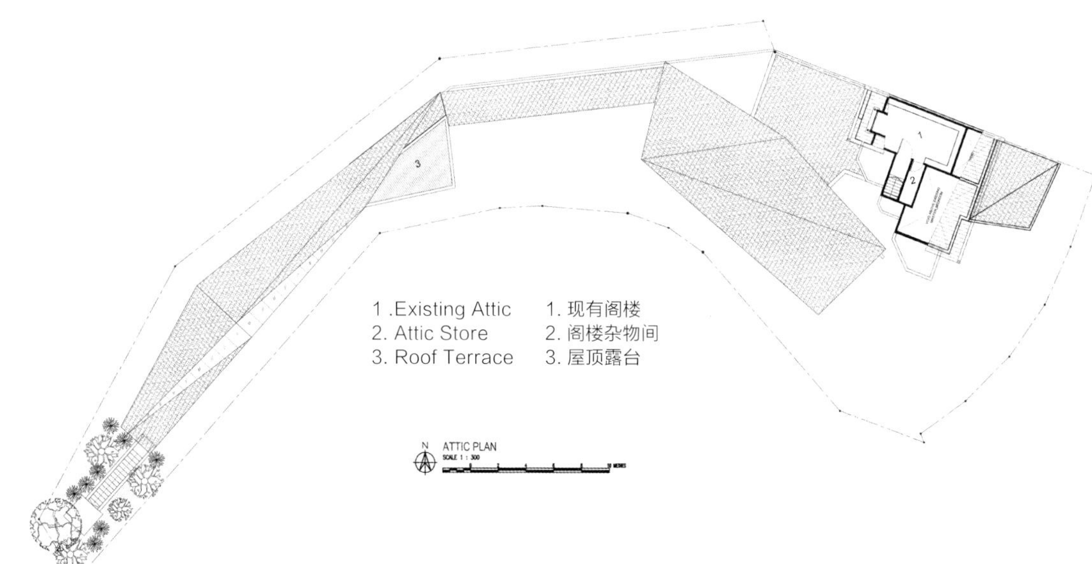

Attic Storey Plan 阁楼层平面图

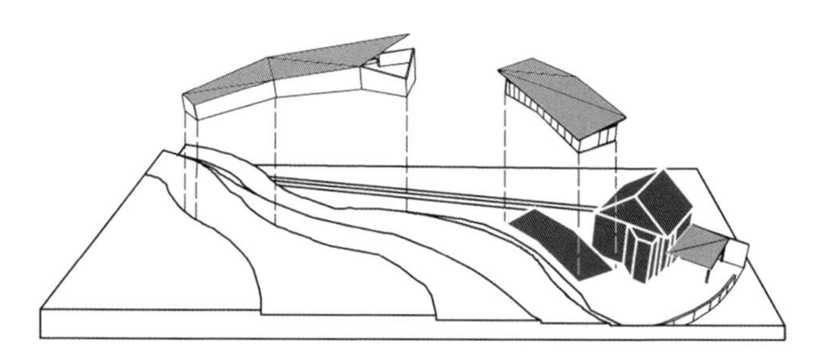

Introduction Of New Pavilion And Rear Extension
新建亭台和后部扩建

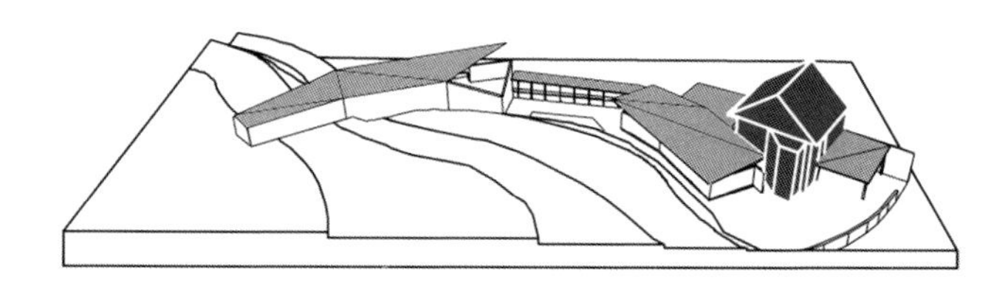

New Linkway Connects Old And New Insertions
连接新旧建筑的走廊

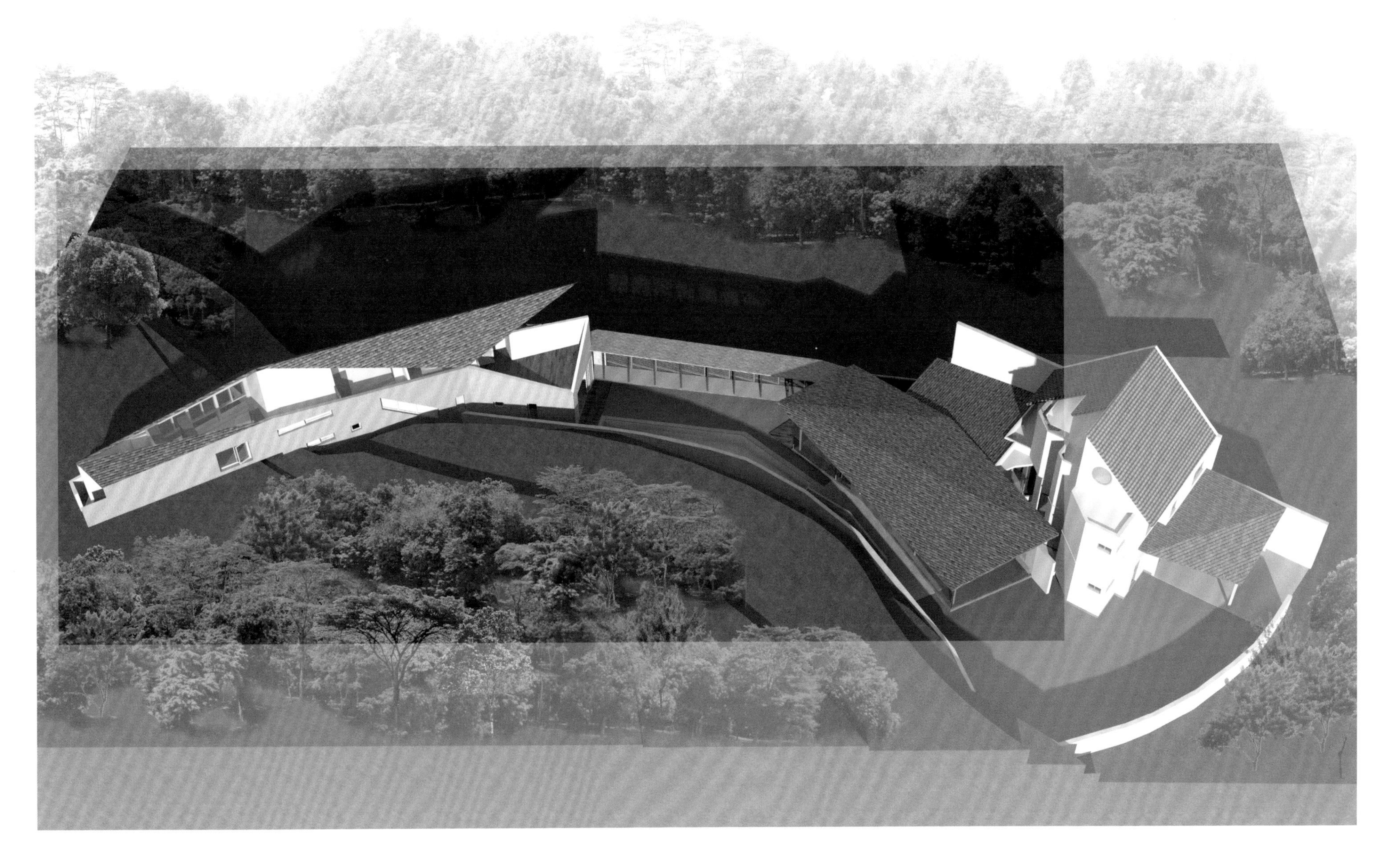

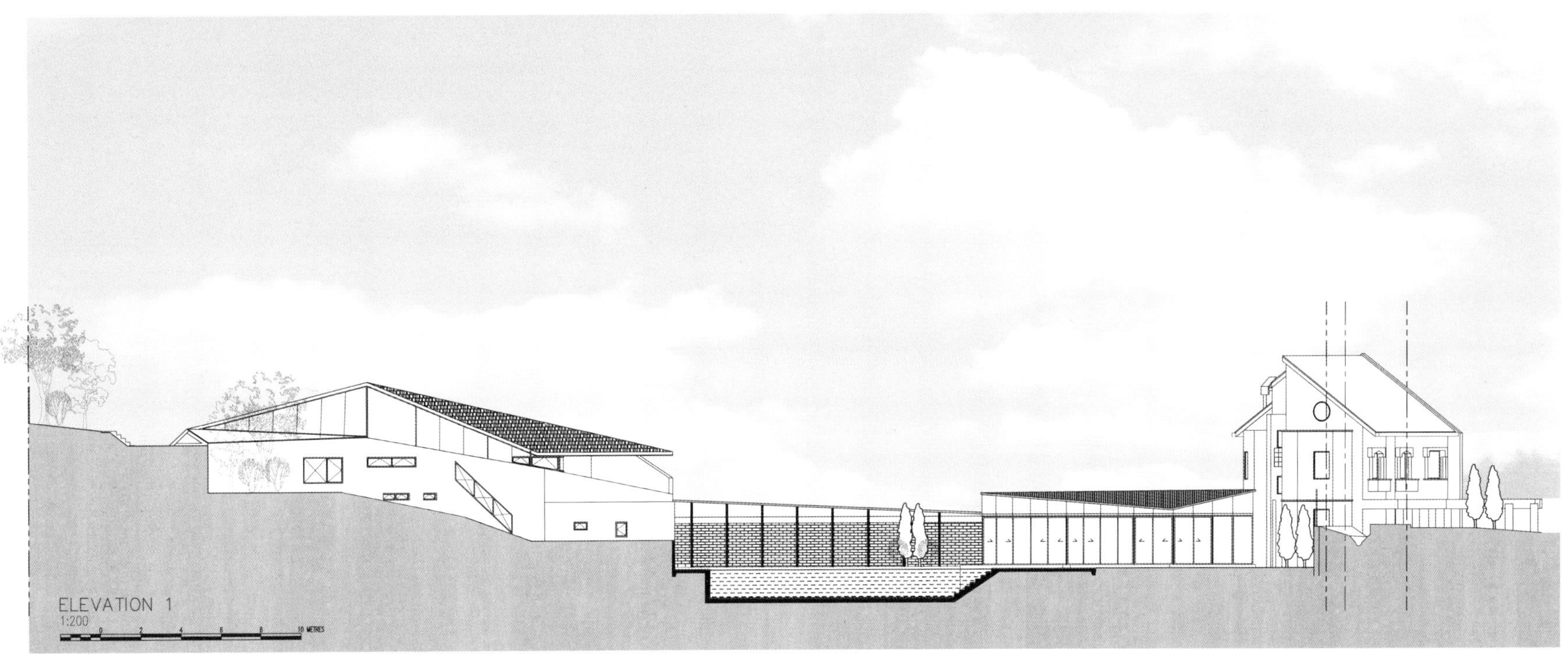

Elevation 1　立面图 1

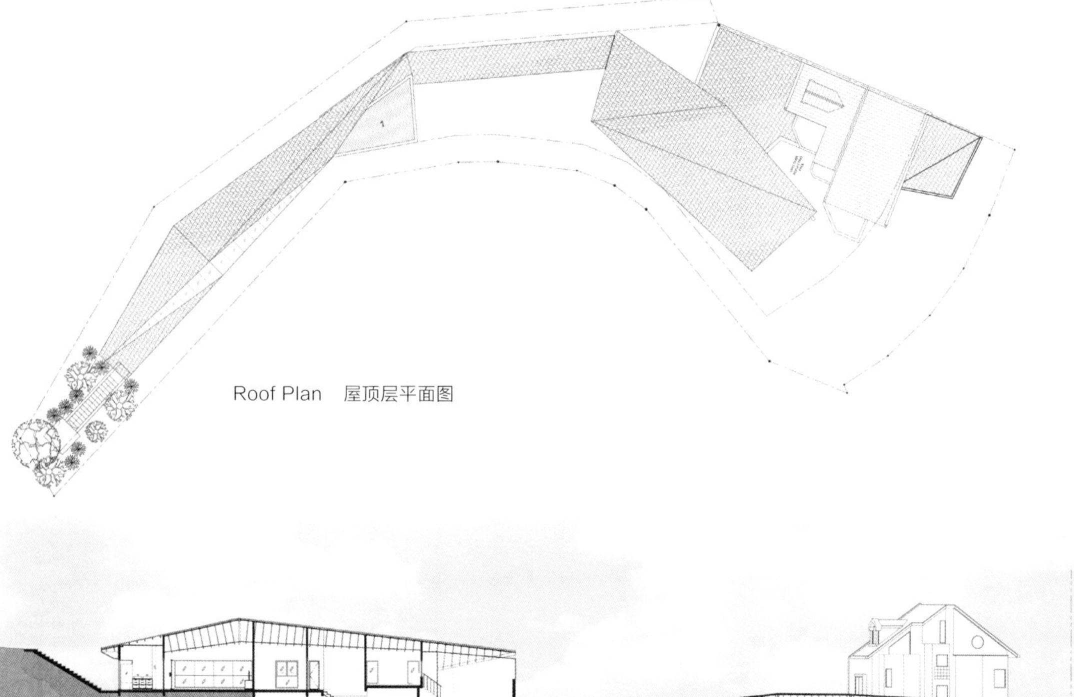

Roof Plan 屋顶层平面图

Section EE 剖面图 EE

项目概况

建筑师们在 Kemunchup 巷别墅改建项目中，面对的是一个极具挑战性的场地且具有非常激动人心的可能性。现保存的半独立式住宅难以觉察地坐落在场地的临街空地。然而，暗藏在这房子后面的是一个 100 m 长的狭窄空地，从属于别墅和向后逐渐缩成一个飞去来器的形状。除了形状狭长，这片土地延伸到基地周围的森林茂密的小山坡，前后斜坡约 6 ~ 7 m。由于这长长的地带被认为是不可建的地区，大部分的场地都被闲置和忽视。10 000 平方英尺（929.03 m^2）的场地外，现有的半独立式住宅只覆盖了大约 30% 的地区。建筑师看到了这个被忽视的地段，尽管是危险地带，却也是场地里的最美丽的部分，建筑师决定挑战自己，在上面建别墅。

建筑设计

为了减少浪费和节省成本，建筑师们决定保留整个两层加阁楼的半独立式住宅，用最小限度的工作量来修补涂料。项目主要集中于场地的后面部分。首先，现有的半独式住宅后面单层扩展区域被拆除，地面梁和平板被重新用于新单层客厅。在别墅的后部第二层阶地上新增一个二层建筑，用游泳池和有顶盖的走廊与客厅隔开，阶梯环绕在四周。由于场地结构要遵循梯田式倾斜的地面形式，特建立了几个屋顶露台供居民享受四周奇妙的景色。为了保持光亮和最小限度的基础工作，新的建筑结构十分低矮和贴地。另一个为减少材料而做的努力就是利用别墅的第五立面——屋顶形式作为向外扩展的主要立面。建筑师们在房子的顶部和两侧利用这种形式弯曲折叠。这种屋顶被认为是热带地区传统倾斜山形墙的屋顶的一种演变，又保留了热带地区屋顶视觉和功能的重要性。从审美的角度上看，这种起伏的屋顶形式也与周围的山丘呼应和有组织地围绕在这块弯曲倾斜的场地四周。

Location: Belimbing Avenue, Singapore

Architectural Design: Hyla Architects

Land Area: 189.2 m^2

Floor Area: 316 m^2

Photography: Derek Swalwell

项目地点：新加坡 Belimbing Avenu

建筑设计：Hyla 建筑事务所

占地面积：189.2 m^2

建筑面积：316 m^2

摄影：Derek Swalwell

Keywords 关键词

Timber Trellis 木框结构

Round Sections 圆形构造

Concrete 混凝土

Belimbing Avenue

Belimbing 街住宅

Overview

A light well with a reflecting pool divides this intermediate terrace house into two blocks. The light well is covered at the top with a glass and timber trellis providing shelter from the rain and yet still allowing light and ventilation to pass through the entire house. The staircase runs along the side and underneath it is a gently cascading water feature which feeds into the pool.

Architectural Design

On the first storey, a screen / display shelf divides the staircase from the living room. The living, dining and dry kitchen occupy the front block and are in one continuous space, whilst the wet kitchen and service areas are in the rear block. A timber bridge crosses the water feature from the first to the second block.

The top of the stair well is also covered in glass and timber. The timber trellis is hung on steel cables giving it a gentle camber. The stairs lead to the roof terrace on the attic level where the master bedroom is. The master bathroom has a similar timber trellis detail over the bath and shower area, but with smaller and round timber sections to fit the smaller scale of the space. The connecting bridges on the second storey and attic level are all not in line with the first level bridge thereby giving a greater sense of space in the light well.

The whole house is in off form concrete and tropical timbers of teak and chengal.

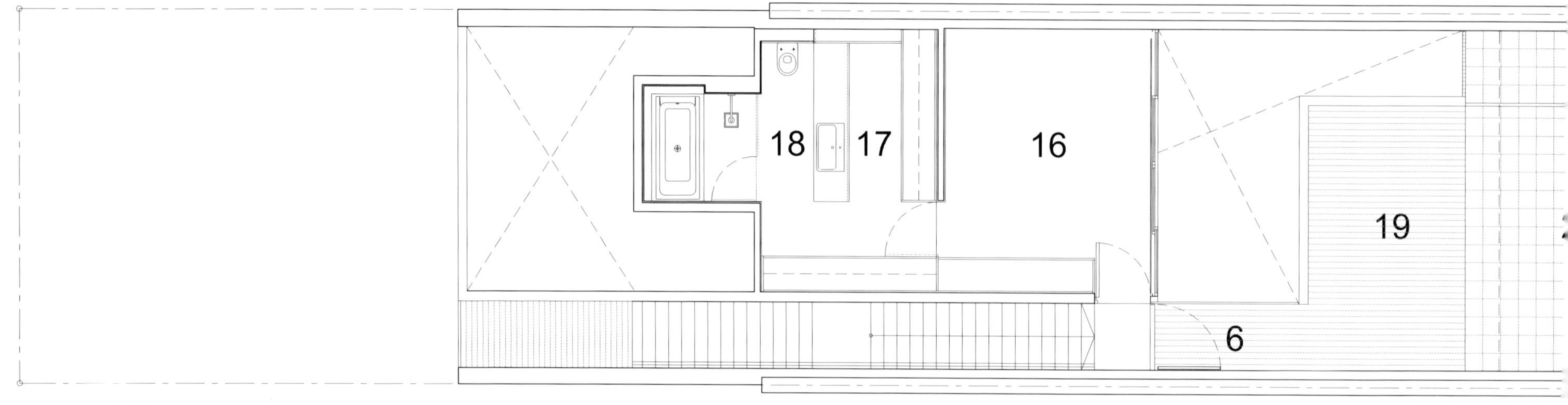

Attic Plan　　阁楼层平面图

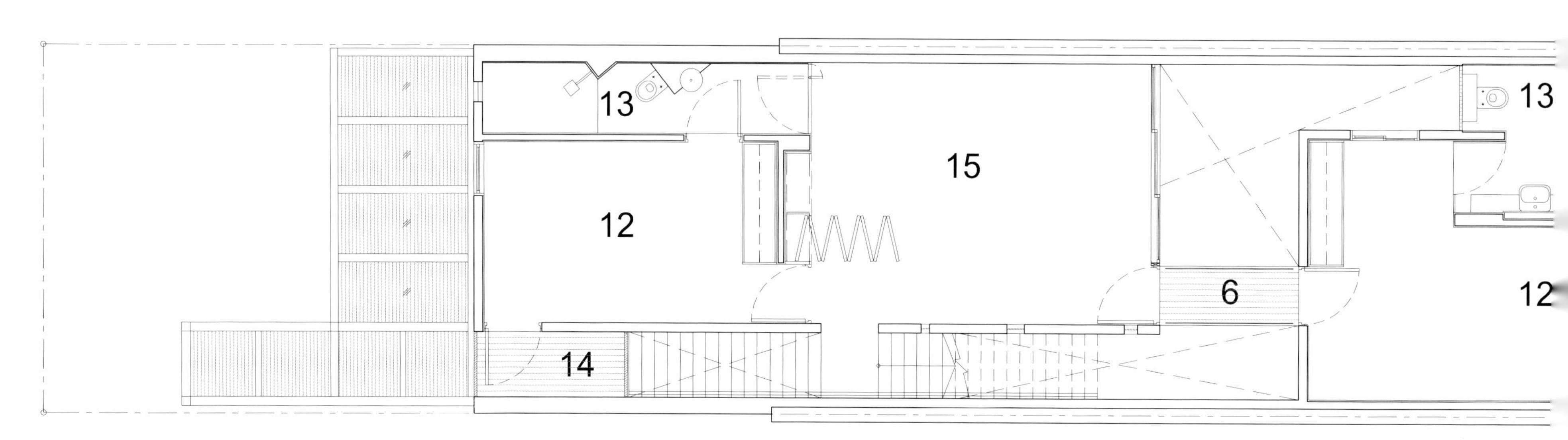

2nd Storey Plan　　二层平面图

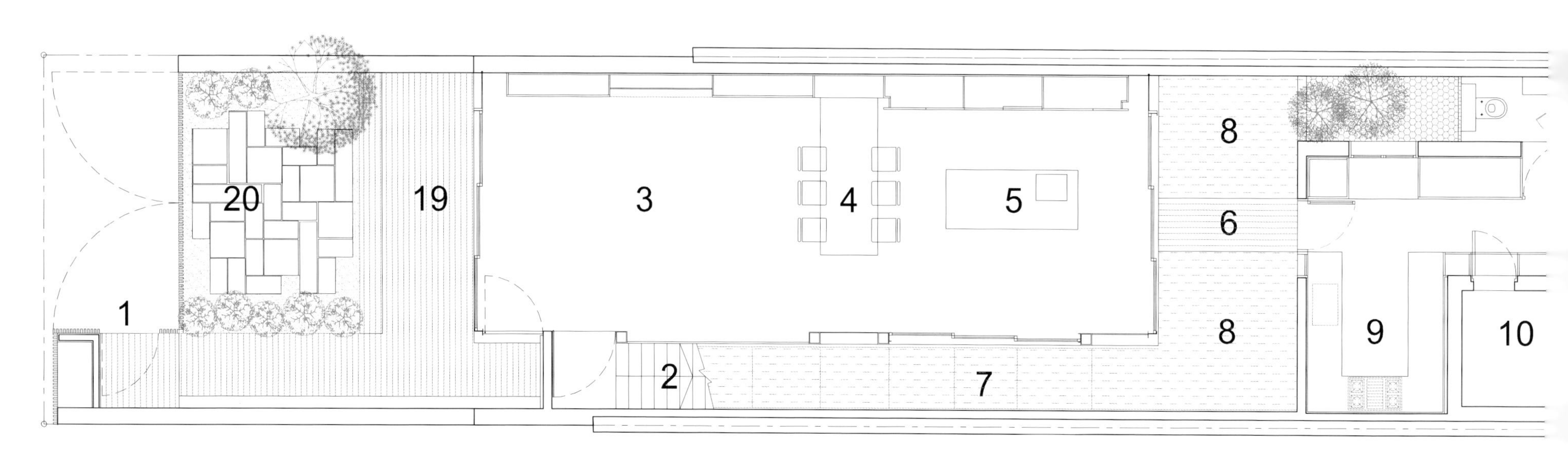

1st Storey Plan　　一层平面图

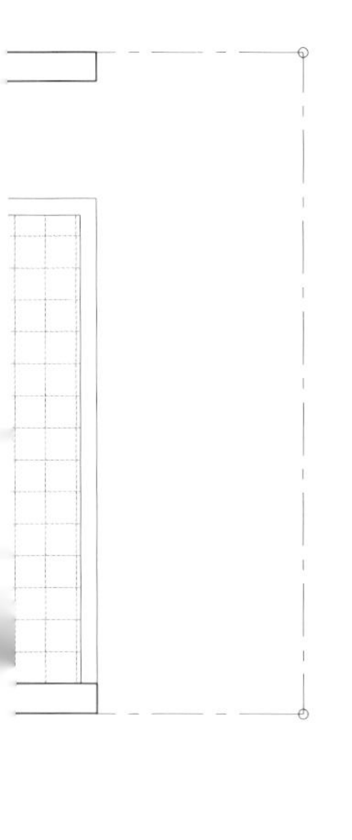

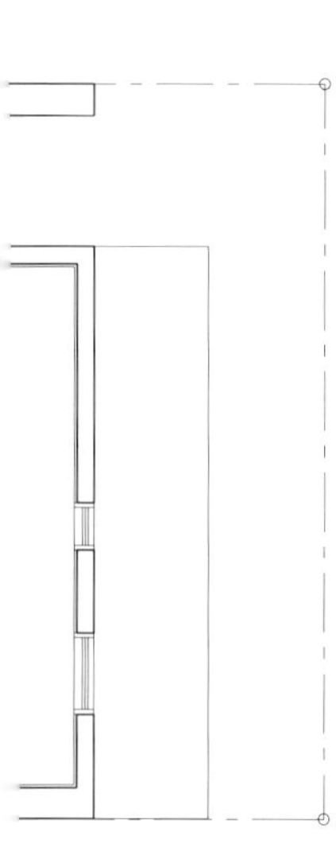

1. Entrance
2. Stairs
3. Living
4. Dining
5. Dry Kitchen
6. Bridge
7. Water Feature
8. Pond
9. Wet Kitchen
10. Shelter
11. Maids
12. Bedroom
13. Bath
14. Balcony
15. Family
16. Master Bed
17. Wardrobe
18. Master Bath
19. Covered Terrace
20. Open Terrace

1. 入口
2. 楼梯
3. 客厅
4. 餐厅
5. 厨房干区
6. 桥梁
7. 水景
8. 池塘
9. 厨房湿区
10. 屏障
11. 佣人房
12. 卧室
13. 浴室
14. 阳台
15. 家庭室
16. 主卧
17. 衣橱
18. 主卧浴室
19. 室内露台
20. 室外露台

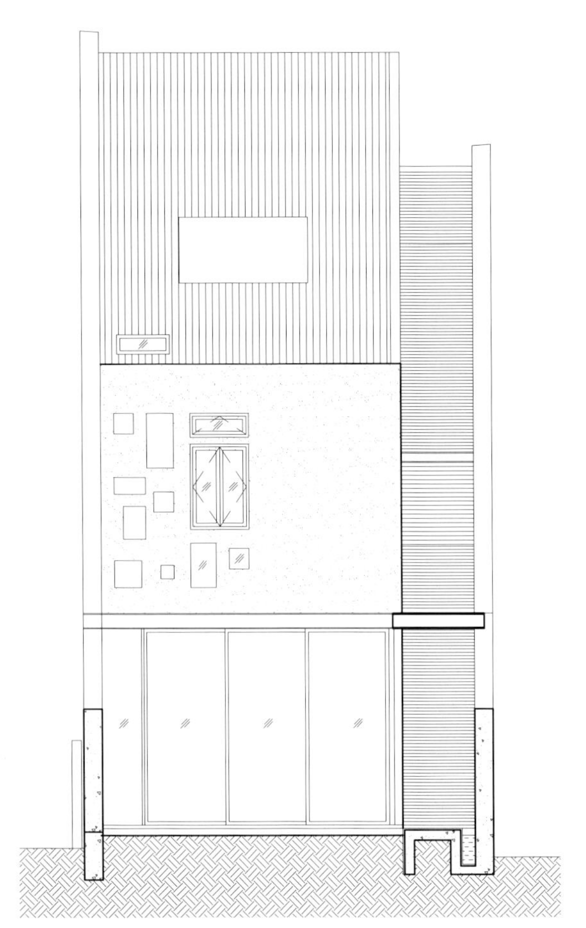

Front Elevation　　正立面图

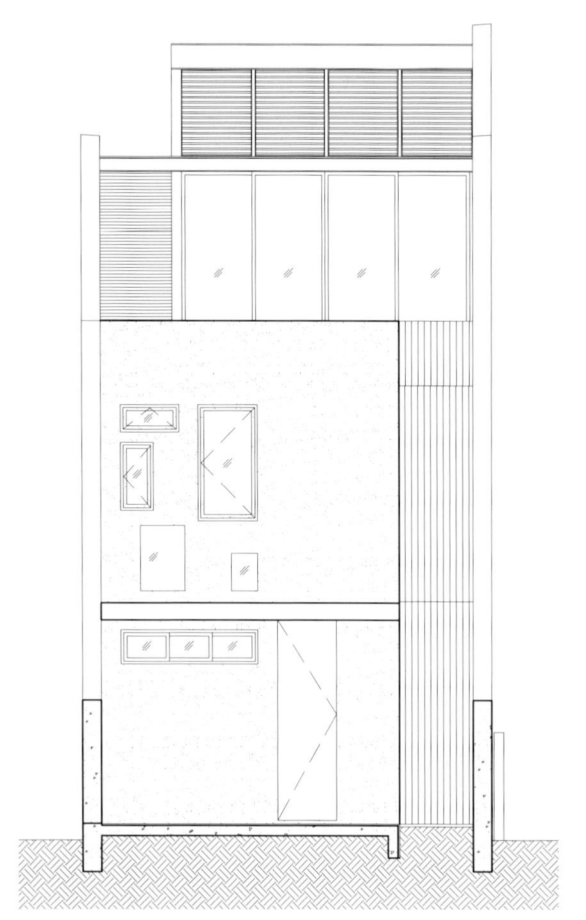

Rear Elevation　　后视图

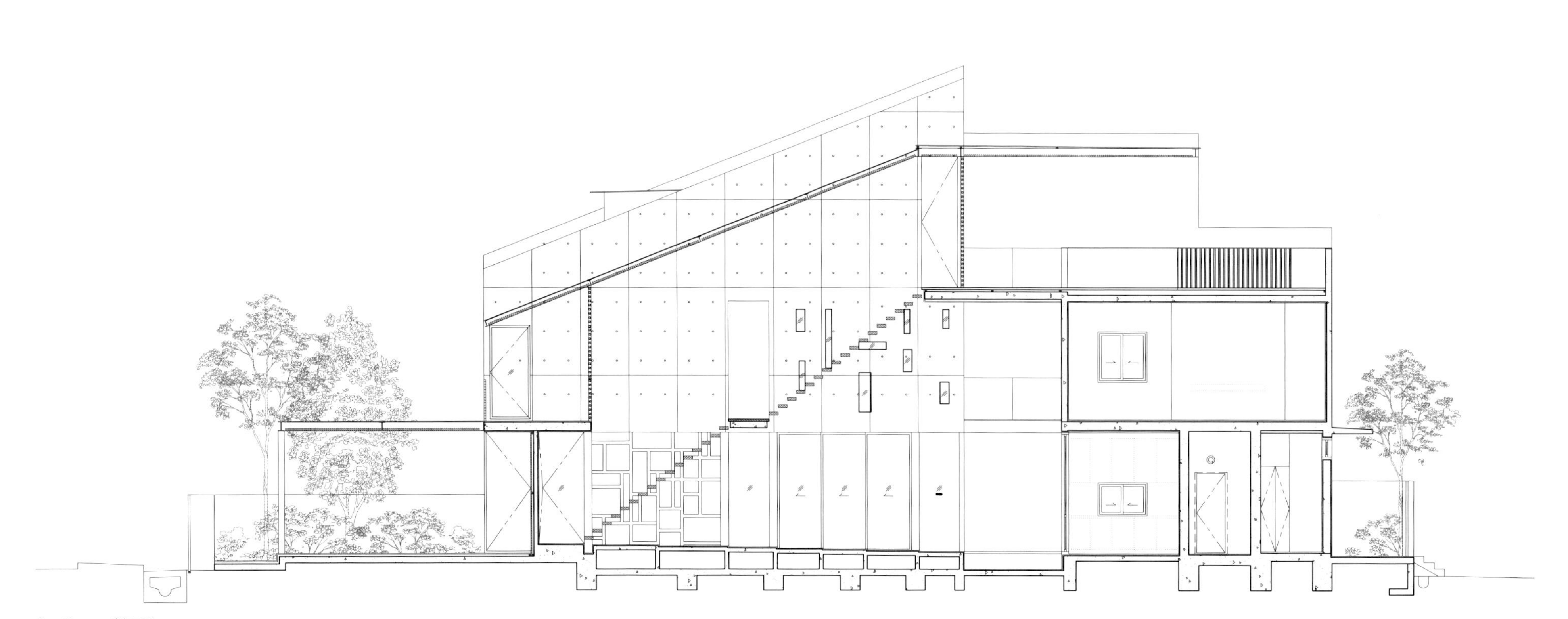

Section　剖面图

项目概况

这座居中的排屋被天井和泳池一分为二。天井的玻璃和木质结构既能遮挡风雨，又能保证室内充足的光照和自然通风。楼梯设在住宅一侧，其下方是雅致的跌水景观，流水直接注入泳池。

建筑设计

一楼设有一座屏风式展架，将楼梯与起居室隔开。客厅、餐厅以及厨房干区位于住宅前部的连续空间中，而厨房湿区和其他的服务区则分布在住宅后部。一座木桥横跨于水景上方，将住宅前后两部分连接起来。

楼梯的顶部同样覆盖着玻璃和木质结构。木框悬挂于钢索上，形成柔和的弧度。楼梯一直通往阁楼上方的屋顶露台，而主卧则设于此处。主卧卫生间的淋浴区上方也设有类似的木框结构，只是受限于空间，采用占地更小的圆形结构。三层空间的连桥并不平行，这样一来，更加强化了天井结构的空间感。

整座建筑就是混凝土结构形式与热带柚木和樟宜树的组合。

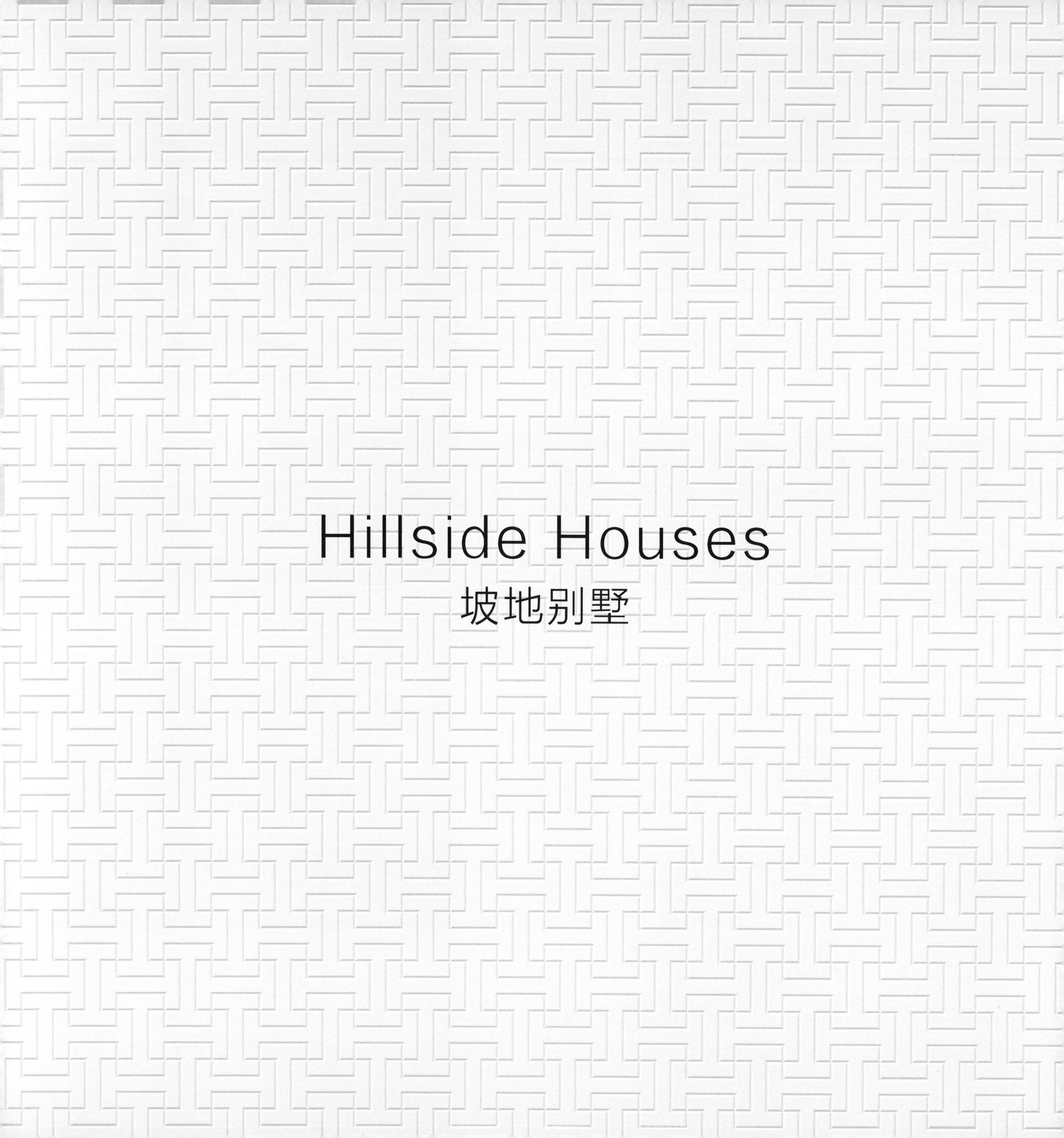

Hillside Houses

坡地别墅

Graceful Form Comfortable Space Compact Layout

形态美观　空间舒适　布局紧凑

Location: Siglap Hill, Singapore

Architectural Design: Aamer Architects

Land Area: 891 m^2

Floor Area: 767 m^2

Photography: Patrick Bingham – Hall

项目地点：新加坡西格莱普

建筑设计：Aamer 建筑事务所

占地面积：891 m^2

建筑面积：767 m^2

摄影：Patrick Bingham - Hall

Keywords 关键词

Curvilinear Plane 曲面造型

Generous Floor 楼层开阔

Big Balconies 宽敞阳台

Siglap
西格莱普住宅

Overview

The house sits on Siglap Hill, the highest point in a residential suburb in the eastern part of Singapore. As the highest point, the site is breezy and it also enjoys spectacular views of the surrounding low-rise neighborhood and the city skyline. The architects designed the house to capitalize on both attributes of the site.

Architectural Design

The main spaces of the house were elevated off the ground to fully exploit the magnificent views and catch the breeze. The main "public" spaces—the living room, dining room and the open kitchen—were placed on the second level while the private spaces—the master bedroom, study and personal entertainment space—were placed on the third level. A single continuous curvilinear plane wraps around the two levels of main spaces, flowing from roof of the car porch to the floor of the second level and upward to join the roof of the third level, unifying the main spaces as a single dynamic volume on the southern edge. The curvilinear plane unwraps itself on the western edge to reveal two big balconies thrusting forward in the direction of the city skyline, propelled by two slanted columns. Around the perimeter of the rooms in the main spaces are streamlined and generously proportioned balconies. These balconies allow activities to spill out from the rooms and also link up the different rooms. As significantly, they also serves as an environmental filter sheltering the rooms from direct sunlight and rain.

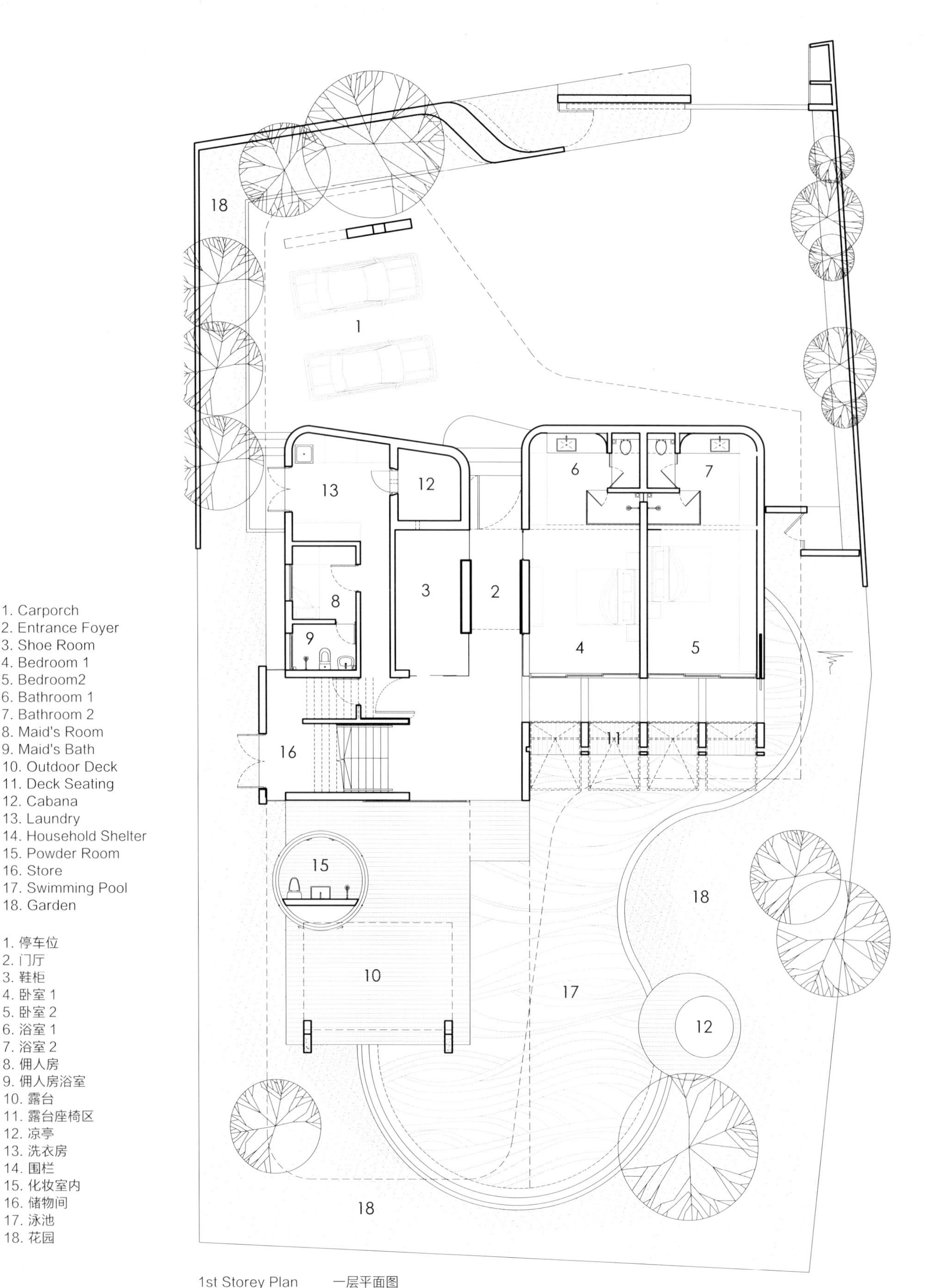

1st Storey Plan　一层平面图

In contrast to the openness of the second and third levels, the ground level appears closed off. From the street entrance, there are two blank walls that defined a recessed entrance. The entrance leads to a swimming pool with a covered terrace on one side. On the other side of the swimming pool are two resort style "Cabana" rooms that are currently used as a gymnasium and a guest bedroom. In contrast to the expansive and extroverted second and third level, the spaces in the first level are rather introverted—the perimeter planting blocks out any outward view and the swimming pool became the inward focus. This contrast is perhaps calculated to accentuate the sense of drama when the guest ascends the staircase and sees the sudden unveiling of the spectacular views.

项目概况

该项目位于新加坡东部郊外住宅区的最高点——西格莱普。基于此，该项目既有良好的通风又能俯瞰周边的低层住宅以及远处城市天际线的迷人美景。建筑师们则充分利用了这些特性。

建筑设计

建筑师抬高该住宅的大部分空间，充分利用迷人美景和良好通风。客厅、餐厅以及开放式厨房，这些主要的公共空间安排在二楼；而私人空间如主卧室、书房和私人娱乐空间则设置在三楼。一条连续的曲面从一楼车库顶延伸出来，经过二楼，最后与三楼的屋顶相连，环绕着二楼和三楼的主要空间。从南面看过来，这些主要空间形成了一个富有活力的卷行形状。这条连续的曲面在住宅的西面并没有连贯起来，使得该住宅有两个面向城市天际线的宽敞阳台。它的主要空间中的房间都是曲线外墙，并且都有宽敞的阳台。这些阳台使住户的活动空间变大，并且将各个房间连接在一起。更重要的是，它们能充当环境过滤器，使房间免受日晒雨淋。

与开放的二楼三楼相比，一楼显得格外的封闭。从街道入口处进来，房子入口嵌在雪白的墙壁之间。一进门便来到了游泳池，泳池一边是走廊，另一边则是两间度假村风格的小屋，一间当做健身室，另一间则布置成了客房。同二楼、三楼的开阔相反，一楼的空间非常的封闭——周边种植的植物遮挡了外来视线，而泳池则成为了中心。这样的反差，也许是为了让来访宾客在上到二楼见到开阔的迷人美景之后，产生更加强烈的对比。

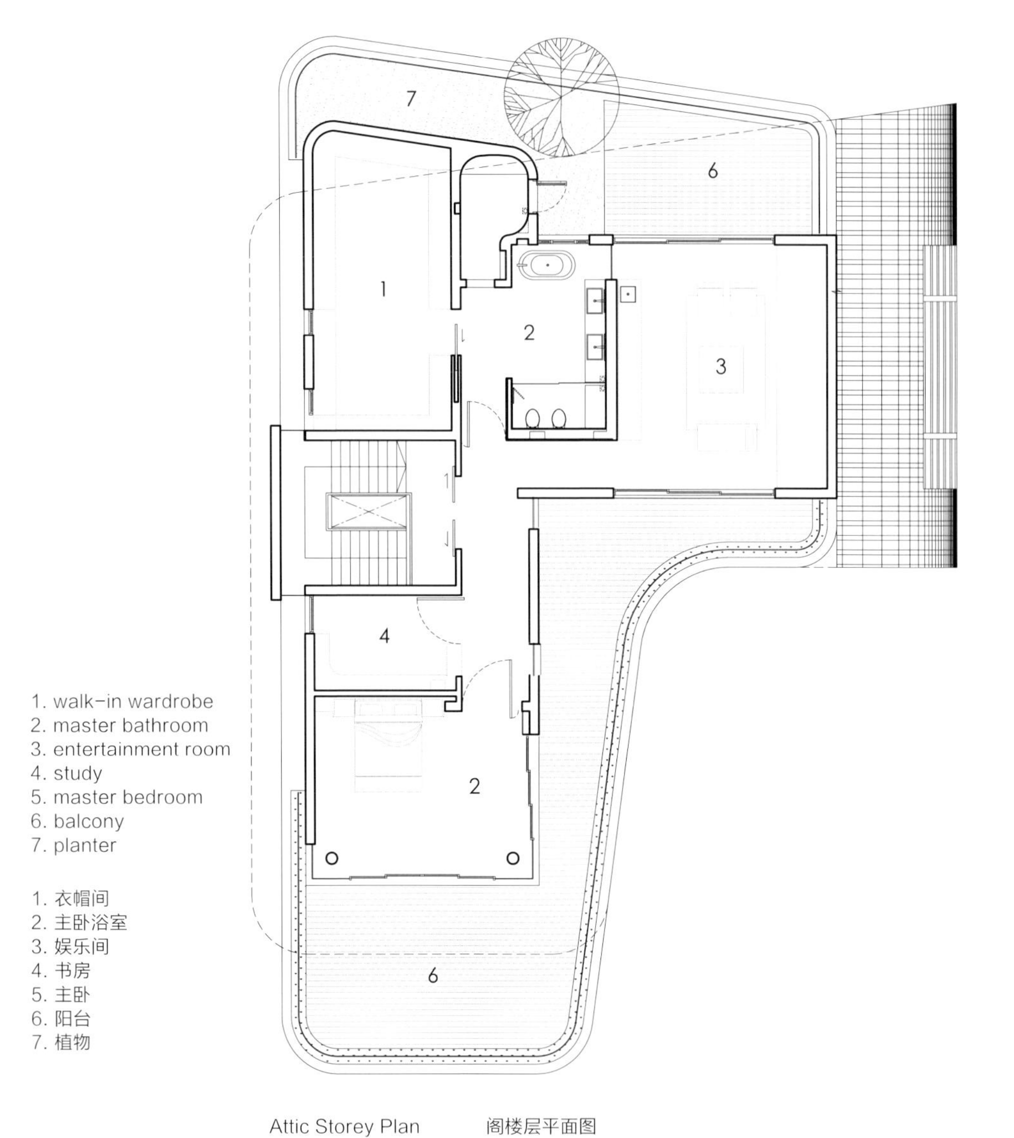

1. walk-in wardrobe
2. master bathroom
3. entertainment room
4. study
5. master bedroom
6. balcony
7. planter

1. 衣帽间
2. 主卧浴室
3. 娱乐间
4. 书房
5. 主卧
6. 阳台
7. 植物

Attic Storey Plan　阁楼层平面图

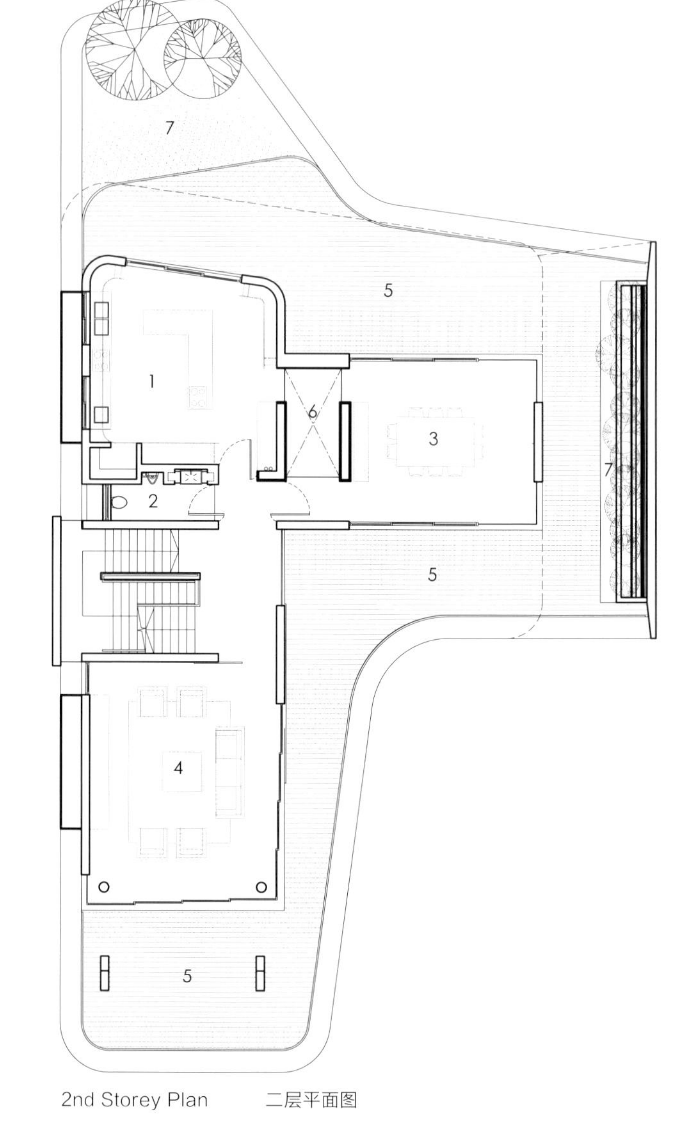

1. kitchen
2. powder room
3. dining
4. living room
5. balcony
6. void
7 .planter

1. 厨房
2. 化妆间
3. 餐厅
4. 客厅
5. 阳台
6. 空地
7. 植物

2nd Storey Plan　二层平面图

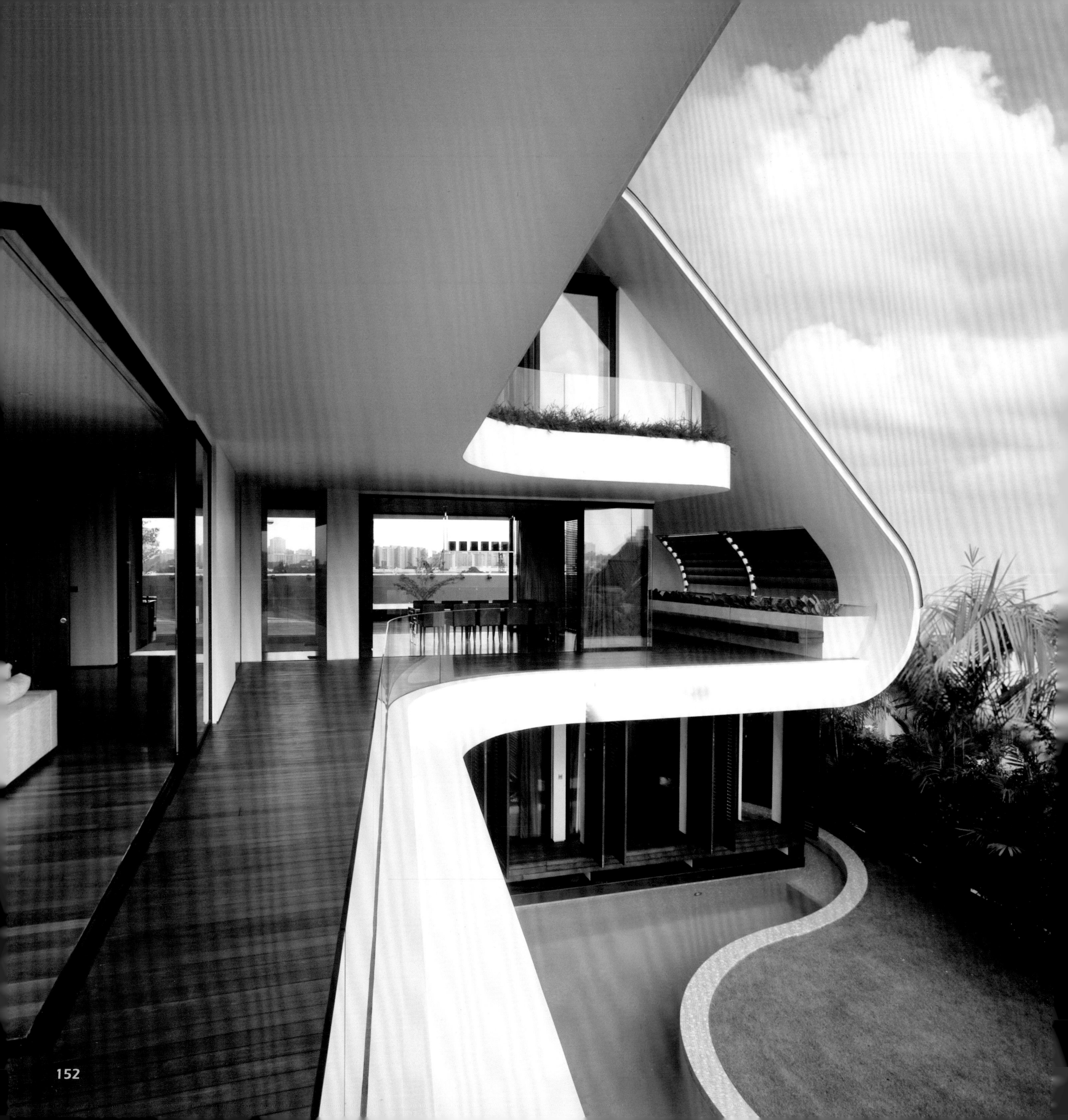

Location: Singapore

Architectural Design: Aamer Architects

Land Area: 1,045 m^2

Floor Area: 882 m^2

Photography: Sanjay Kewlani

项目地点：新加坡

建筑设计：Aamer 建筑事务所

占地面积：1 045 m^2

建筑面积：882 m^2

摄影：Sanjay Kewlani

Keywords 关键词

Long Verandah Bridge 狭长走廊

Outdoor Swimming Pool 户外泳池

Private Space 私密空间

Maryland Drive

Maryland 住宅

Overview

The site is split into two levels with about 1.5m drop from the entrance level and the home is designed as two blocks, with a master and a children's block, linked by a verandah bridge over the long swimming pool.

Architectural Design

Large covered verandahs extend the interior spaces to the large outdoor poolside areas, i.e., creating that charming colonial lifestyle of a bygone era.

A large garden with lush landscape complements the tropical living environment. The cute 'out-house' is meant to provide for some privacy space where one could quietly reflect on daily life's ups and downs and thereafter obtain inner peace.

项目概况

该项目的入口和房子分别在两个层级，高差约 1.5 m。房子建有两栋，分别为主人房和儿童房，由一个跨过狭长泳池的走廊相连。

建筑设计

这个走廊连接了室内空间和户外宽敞的泳池，为住户营造出一种仿佛生活在旧时代的错觉。

花园里大量的景观设计增加了住户在热带地区生活的舒适感。该住宅中有一座与两栋房分离的小房间，是为了能给住户提供更多隐私空间，令住户能够安静地思考日常生活的种种，进而达到内心的宁静。

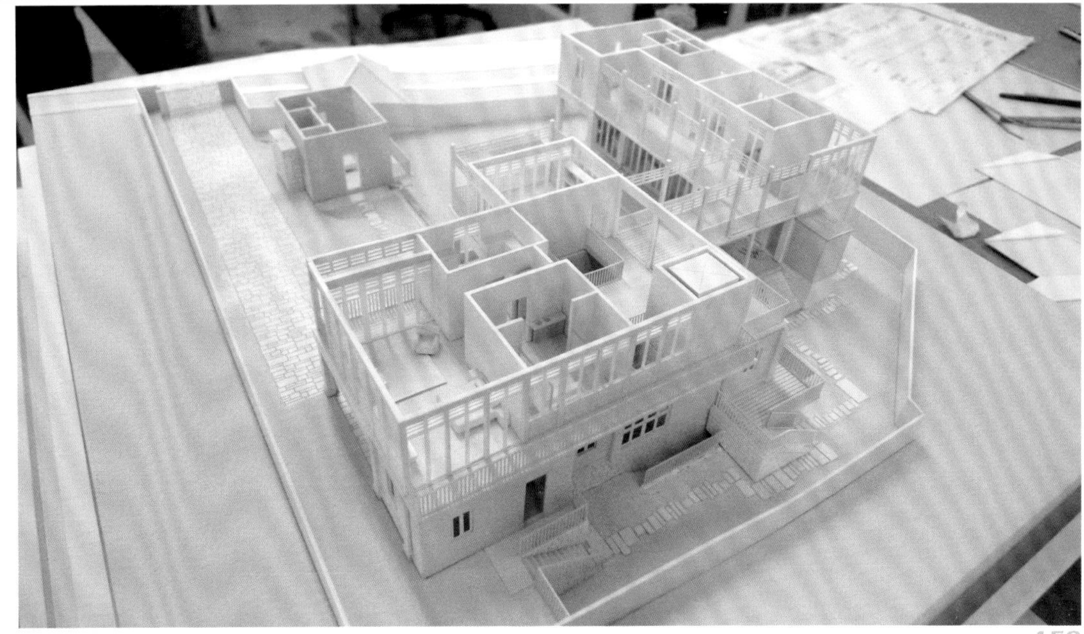

TIBHAR

Location: 25 Olive Road, Singapore

Architectural Design : AKTA

Interior Design Consultant : 7 Interior Architecture

Total Area : 929.03 m^2

项目地点：新加坡奥利夫路 25 号

建筑设计：AKTA

室内设计：7 Interior Architecture

总建筑面积：929.03 m^2

Keywords 关键词

Refined Materials 选材精心

Luxurious Experience 奢华体验

Inherent Structure 结构完善

25 Olive Road

奥利夫路 25 号住宅

Overview

Nestled on a 15,800 ft^2 (about 1,467.86 m^2)plot along the winding flank of Caldecott hill, the house overlooks the polo club.

Interior Design

The interior design concept was to refine and harmonize the finishes in the bungalow with her natural setting. Maintaining the inherent structure and layout of the house, the material scheme was deliberated to manifest the nature fronting the house with a textured and luxurious experience.

项目概况

该住宅位于加利谷山蜿蜒的山坡上，占地面积 15 800 ft^2（约 1 467.86 m^2），可俯瞰山脚下的马球俱乐部。

室内设计

在保持住宅内部结构和空间布局的前提下，设计师精心选材，突出屋前的自然景观，营造了充满质感的奢华体验。室内设计旨在完善小屋设计，并使之与周围的自然环境完美融合。

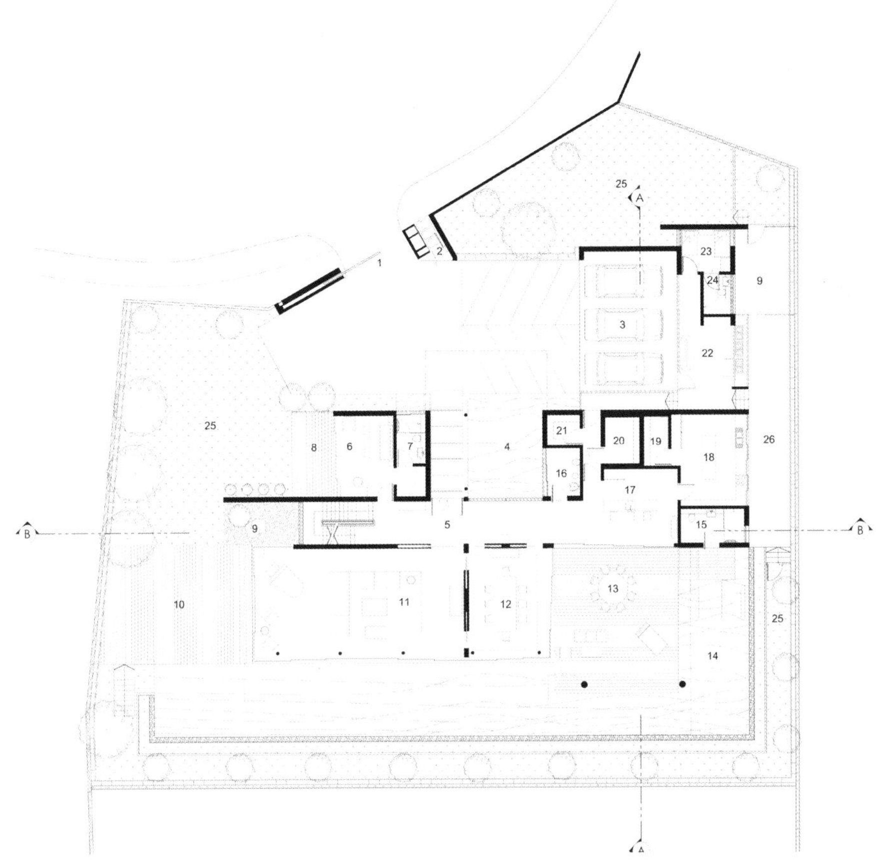

1st Storey Plan　一层平面图

1. Driveway Gate	1. 车道入口
2. Pedestrian Side Gate	2. 人行入口
3. Car Porch	3. 车库
4. Refective Pond	4. 镜面池塘
5. Entrance Foyer	5. 入口门厅
6. Study/Guest Room	6. 书房 / 客房
7. Bath Room	7. 浴室
8. Covered Terrace	8. 有顶露台
9. Dry Landscape	9. 干景观
10. Open Terrace	10 室外露台
11. Living Area	11. 客厅区
12. Dining Area	12. 餐厅区
13. Outdoor Dining	13. 室外餐饮区
14. Lap Pool	14. 小型泳池
15. Bathroom	15. 浴室
16. Powder Room	16. 化妆间
17. Dry Kitchen	17. 厨房干区
18. Wet Kitchen	18. 厨房湿区
19. Larder	19. 食品柜
20. Wine Cellar	20 酒窖
21. Store	21. 储物间
22. Laundry Yard	22. 洗衣房
23. Maids' Bedroom	23. 佣人房
24. Maids' Bathroom	24. 佣人浴室
25. Landscaped Garden	25. 景观花园
26. Service Yard	26. 杂作坊

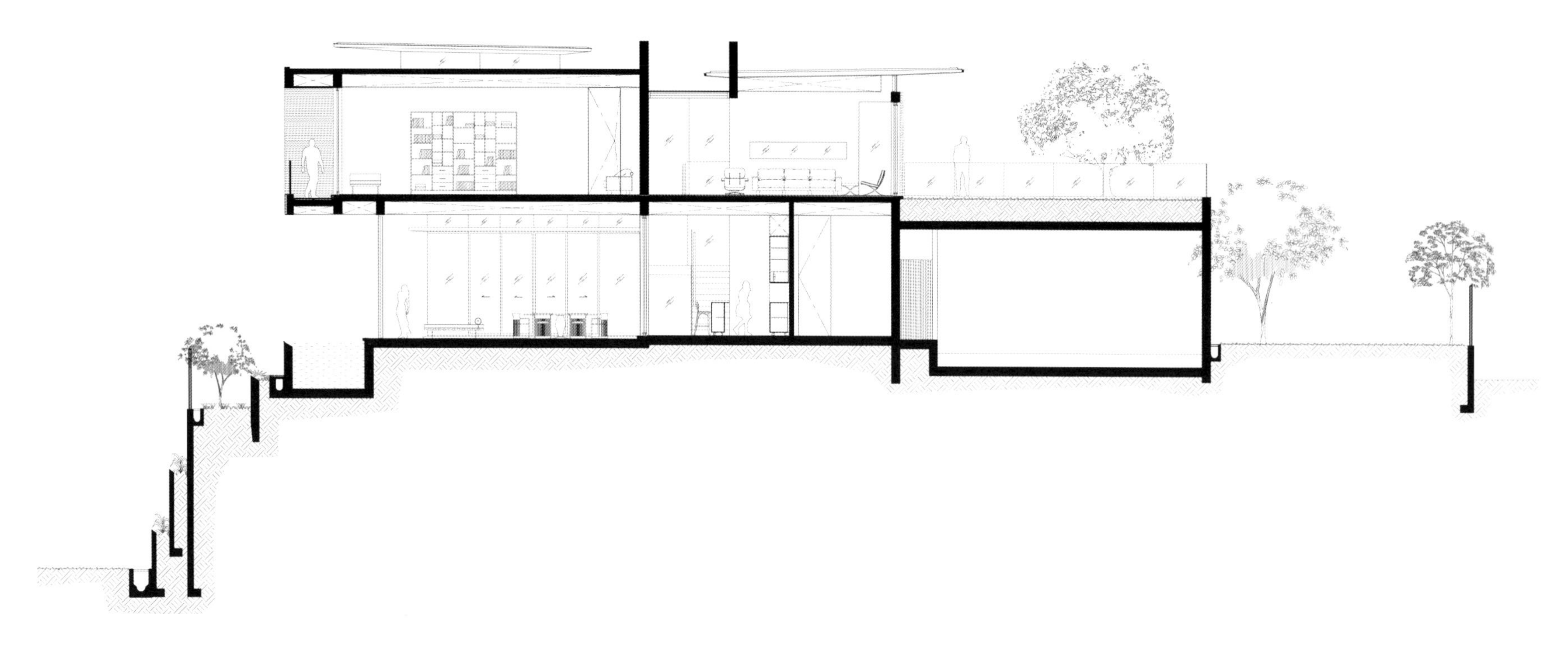

Sections AA　　剖面图 AA

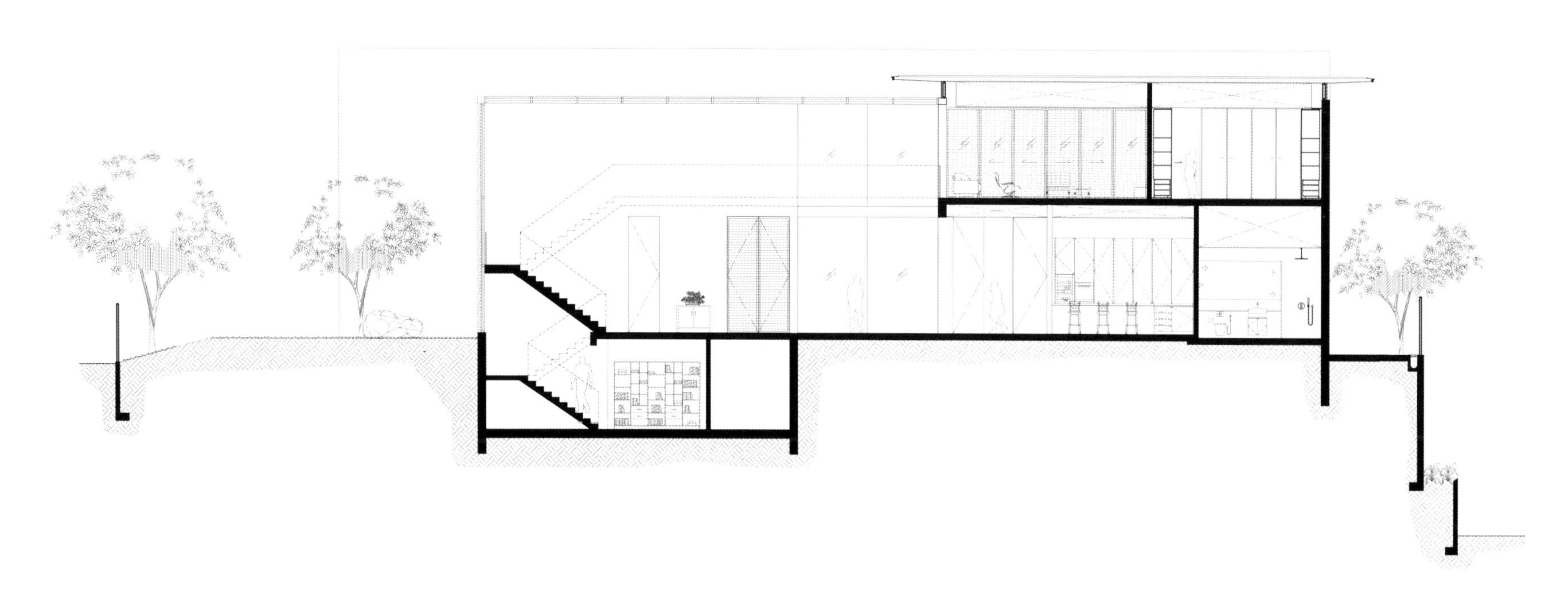

Sections CC　　剖面图 CC

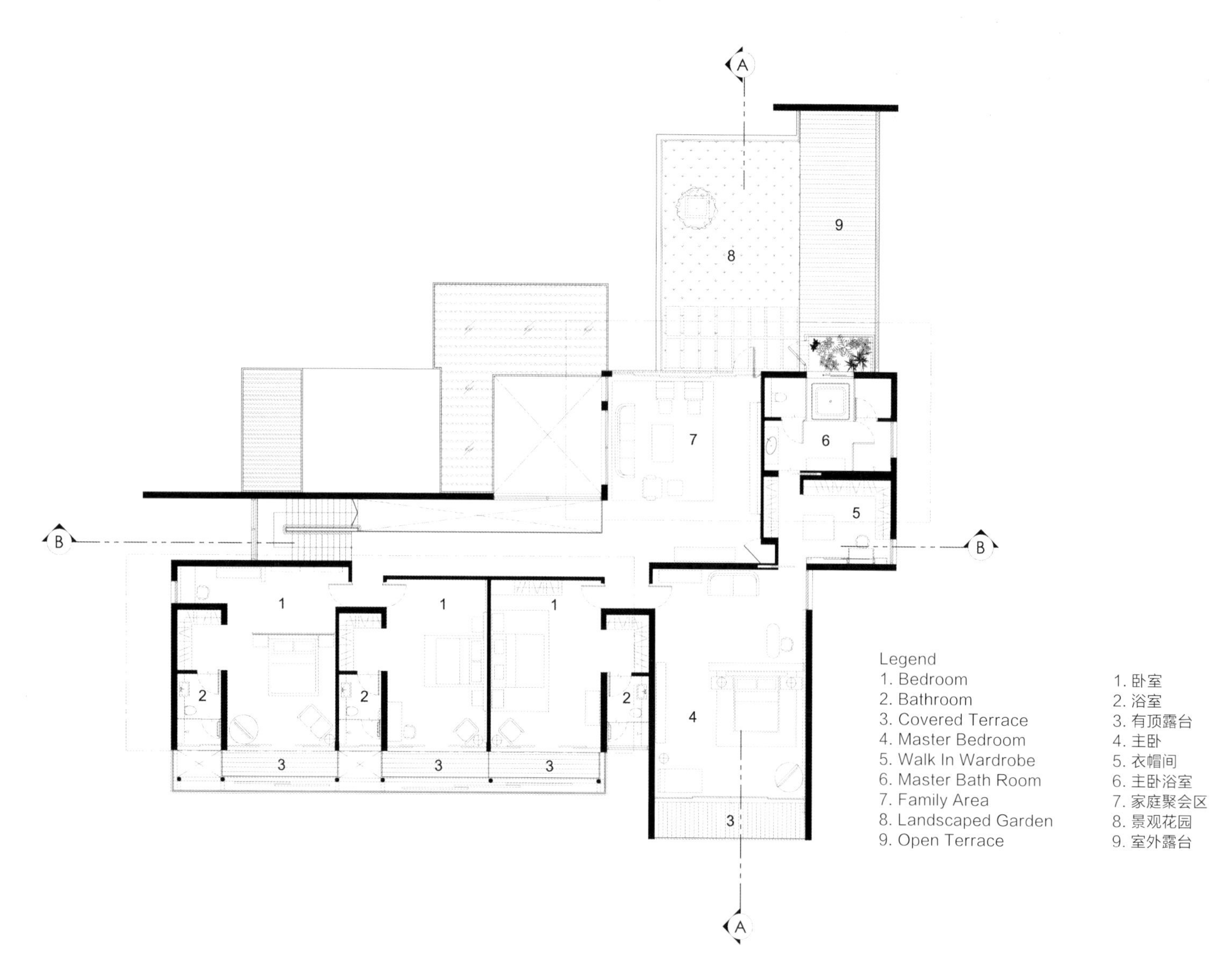

2nd Storey Plan　二层平面图

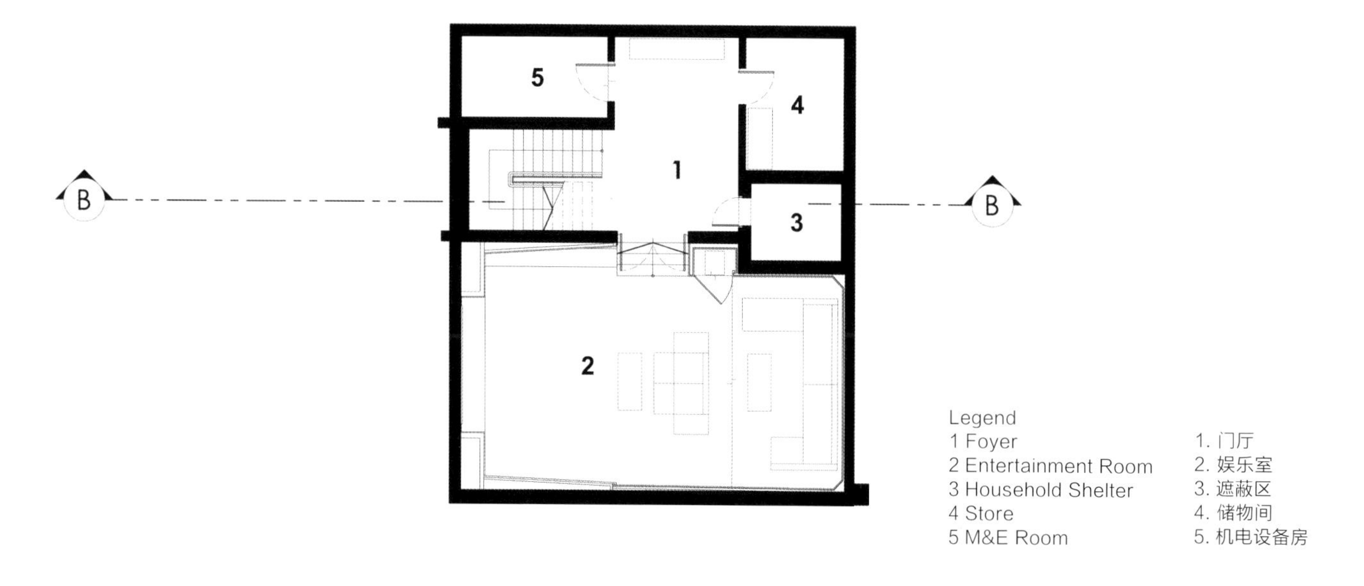

Basement Plan　地下层平面图

Location: Andrew Road, Singapore

Architectural Design: A D Lab Pte Ltd

Design Team: Warren Liu, Lim Pin Jie

Land Area: 1,000m^2

Floor Area: 600m^2

Photography: Derek Swalwell

项目地点：新加坡安德鲁路

建筑设计：新加坡 A D Lab 有限公司

设计团队：Warren Liu，Lim Pin Jie

占地面积：1 000m^2

建筑面积：600m^2

摄影：Derek Swalwell

Keywords 关键词

Sunken Court 下沉庭院

Upward and Downward 层层递进

Privacy 私密性

Andrew Road

安德鲁路别墅

Overview

Situated in the well-known Caldecott region of central Singapore, this bungalow by A D Lab enjoys the relaxed atmosphere of its quaint residential neighborhood with gorgeous views of the MacRitchie Reservoir, one of the nation's most popular nature reserves. The architects, however, had to contend with the site's proximity to a busy highway, that although is not visible from the site, creates a significant amount of noise.

Architectural Design

The undulating terrain of the neighborhood is quite unusual in Singapore, and creates a streetscape whereby the plots of land are about a storey below the street level. As such, each house along the street is entered from the second storey level. The architects took advantage of this unusual situation to lower most of the communal facilities down into a sunken court that shields them from the noise of the highway, as well as heightens the level of privacy and intimacy of the house. From the external entrance of the house, the architectural expression is very understated. The architects kept the built form above street level as ground-hugging as possible by making the roofs appear as folds and peels in the landscape. This understated expression is further assisted by the folding of the roof downward toward the outer edges of the house. The internal spatial expression of this tilting roof form gives the opposite experience, whereby the entrance to the spaces are low and rise upward from the entry, creating the sensation of an enlarging, grand internal room that simultaneously leads the eyes upward to the sky, as well as downward to the intimate central courtyard below. The undulating and folding roof form is covered in turf, further defining it as a continuation of the surrounding green landscape. At certain locations, the roof folds up from the earth level to allow wind and light into the sunken courtyard.

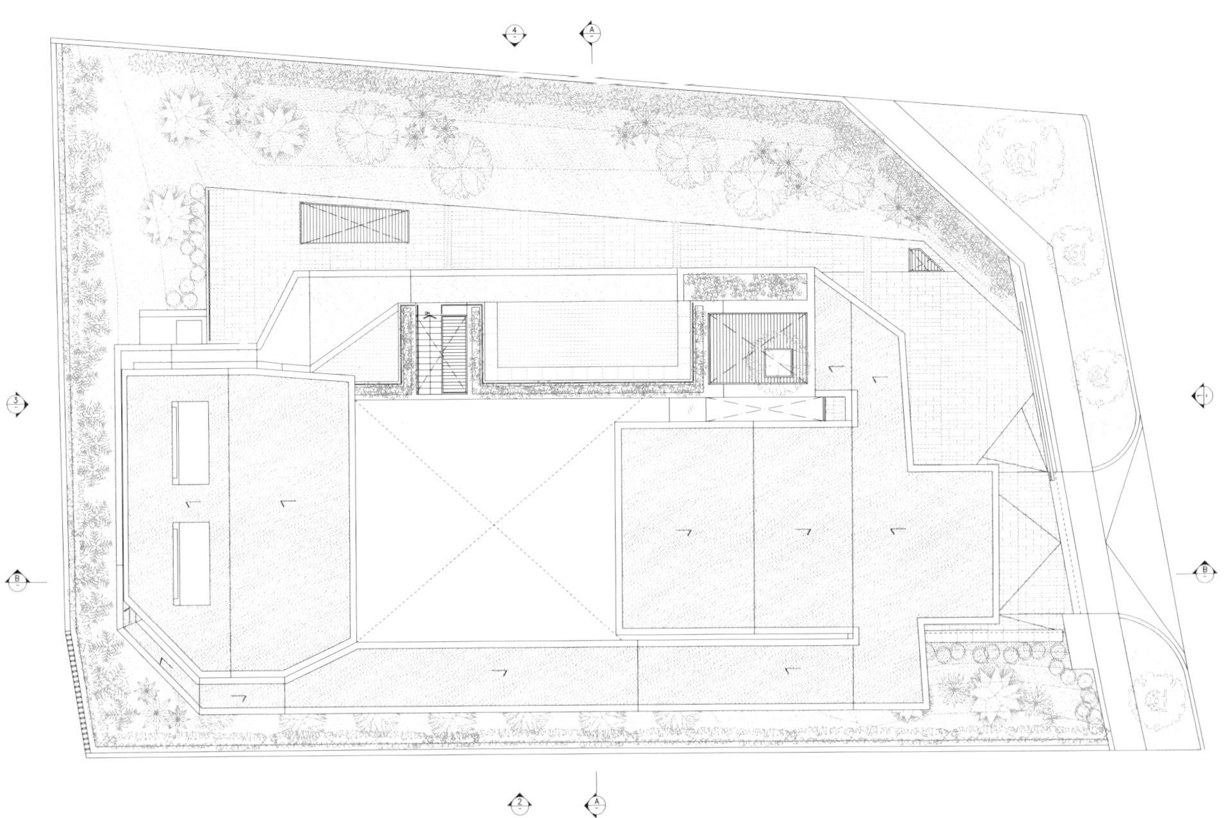

Site Plan　总平面图

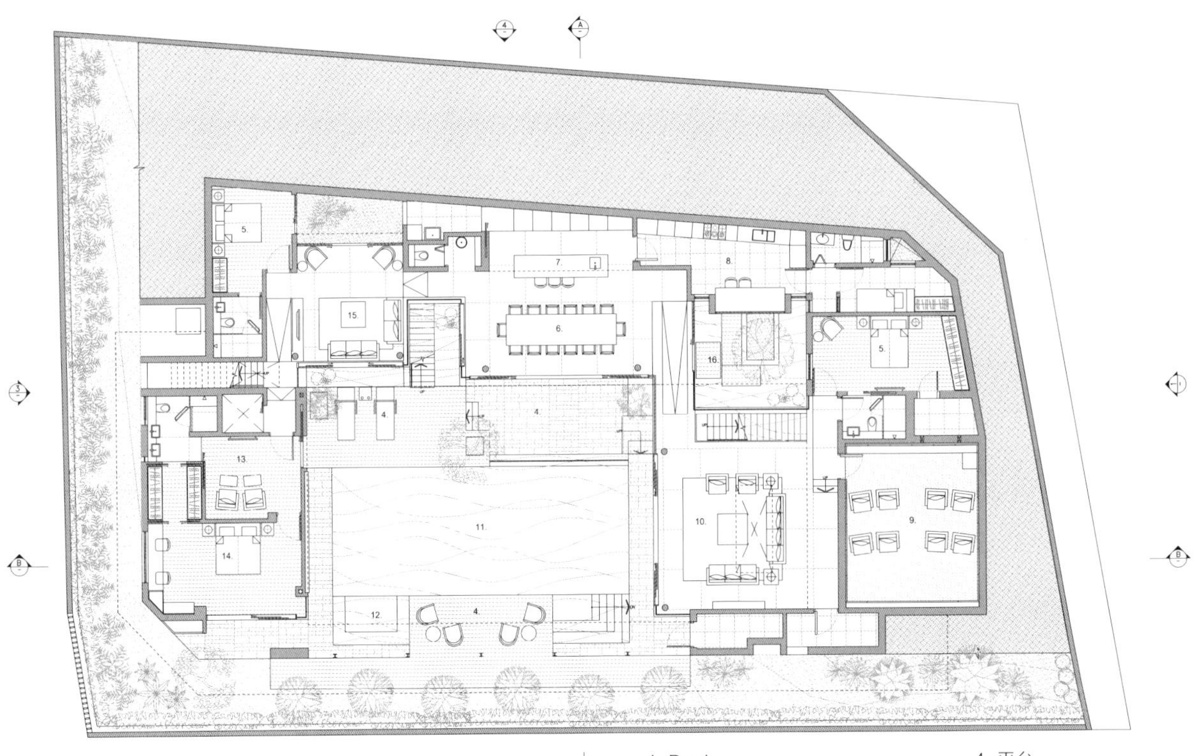

1st Floor Plan　一层平面图

4. Deck
5. Bedroom
6. Dining Room
7. Dry Kitchen
8. Wet Kitchen
9. TV Room
10. Living Room
11. Swimming Pool
12. Jacuzzi
13. Junior Master Suite Lounge
14. Junior Master Suite
15. Family Room
16. Courtyard

4. 平台
5. 卧室
6. 餐厅
7. 厨房干区
8. 厨房湿区
9. 电视厅
10. 客厅
11. 泳池
12. 按摩池
13. 小主人套房休息室
14. 小主人套房
15. 家庭室
16. 庭院

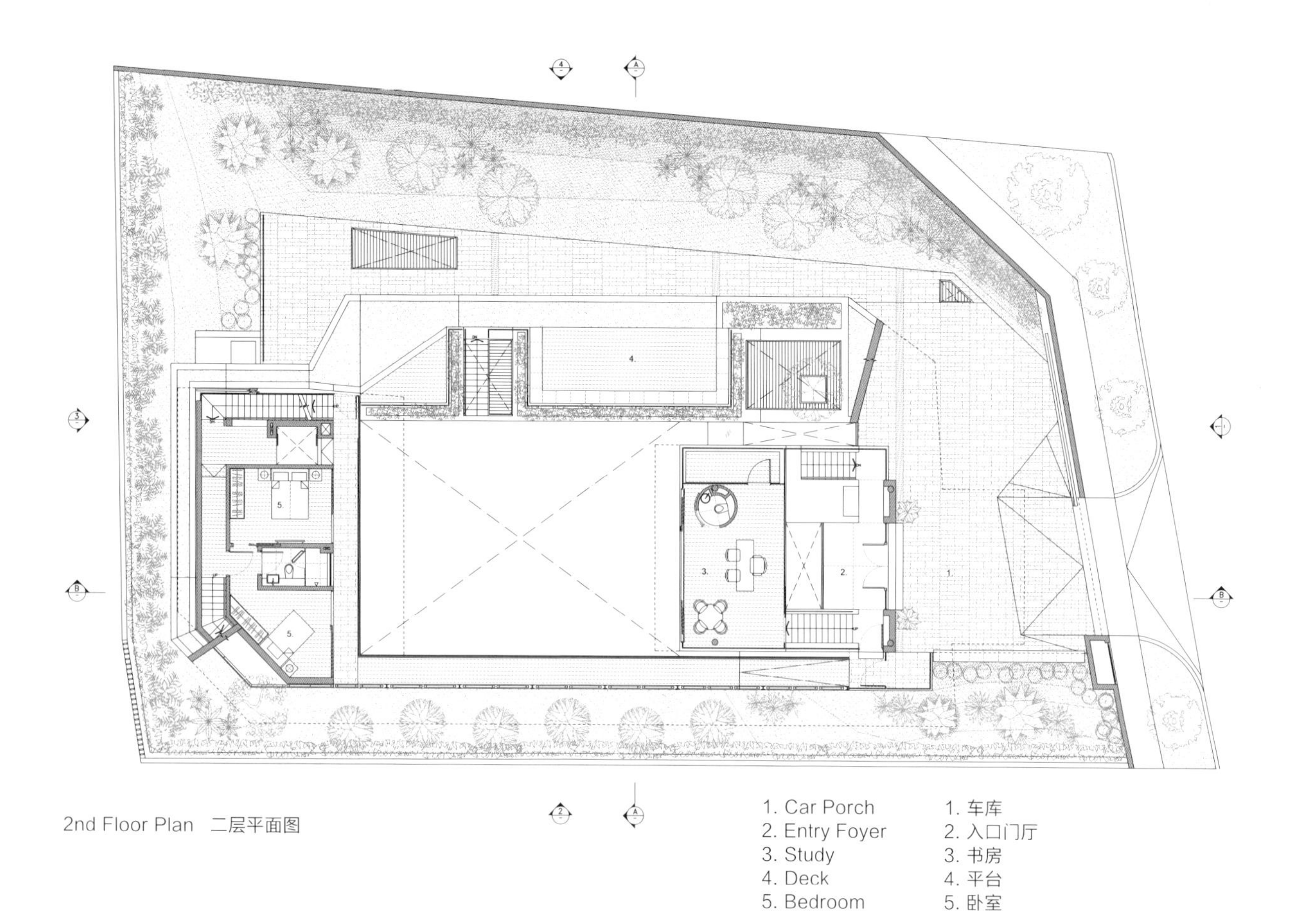

2nd Floor Plan 二层平面图

1. Car Porch 1. 车库
2. Entry Foyer 2. 入口门厅
3. Study 3. 书房
4. Deck 4. 平台
5. Bedroom 5. 卧室

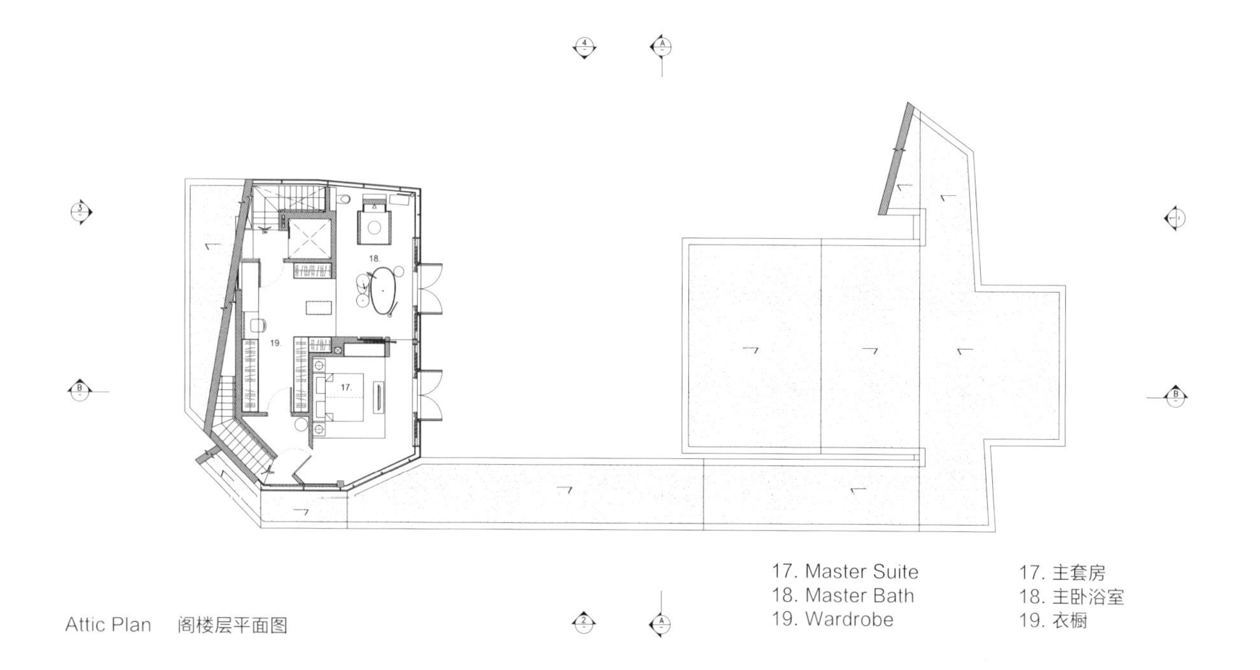

Attic Plan 阁楼层平面图

17. Master Suite 17. 主套房
18. Master Bath 18. 主卧浴室
19. Wardrobe 19. 衣橱

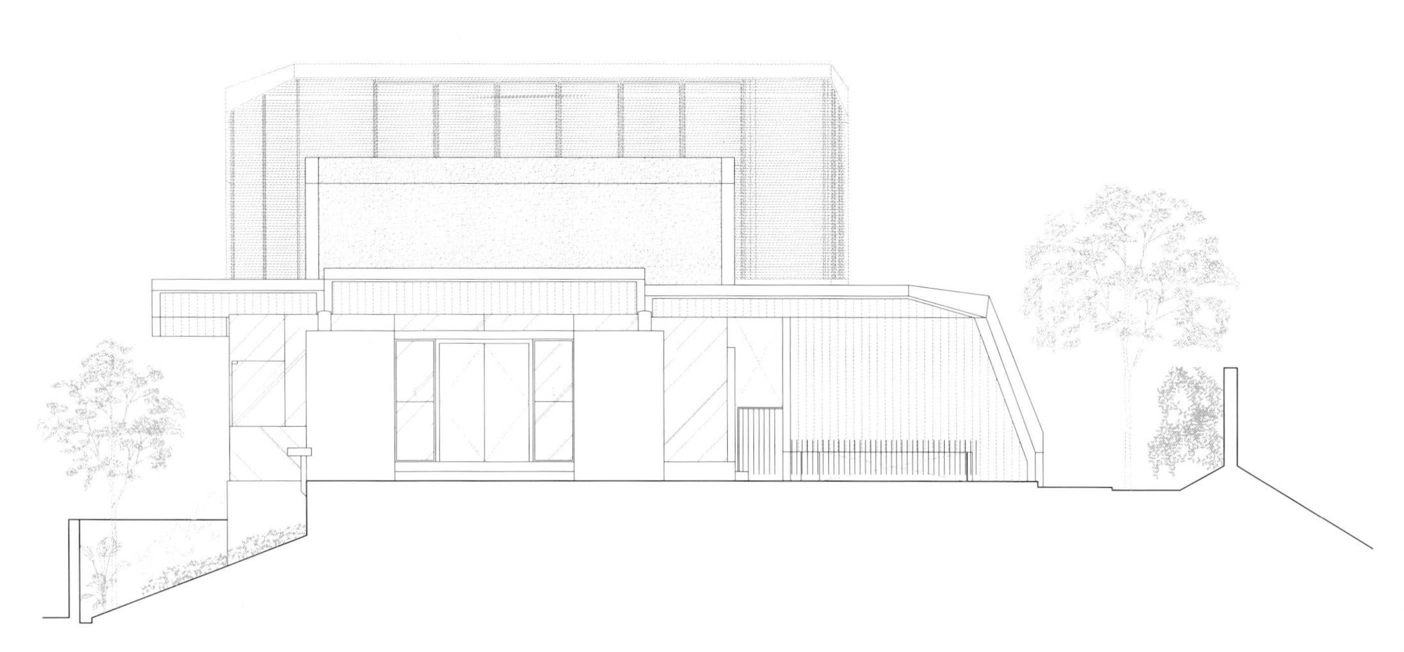

Elevation 01　立面图 1

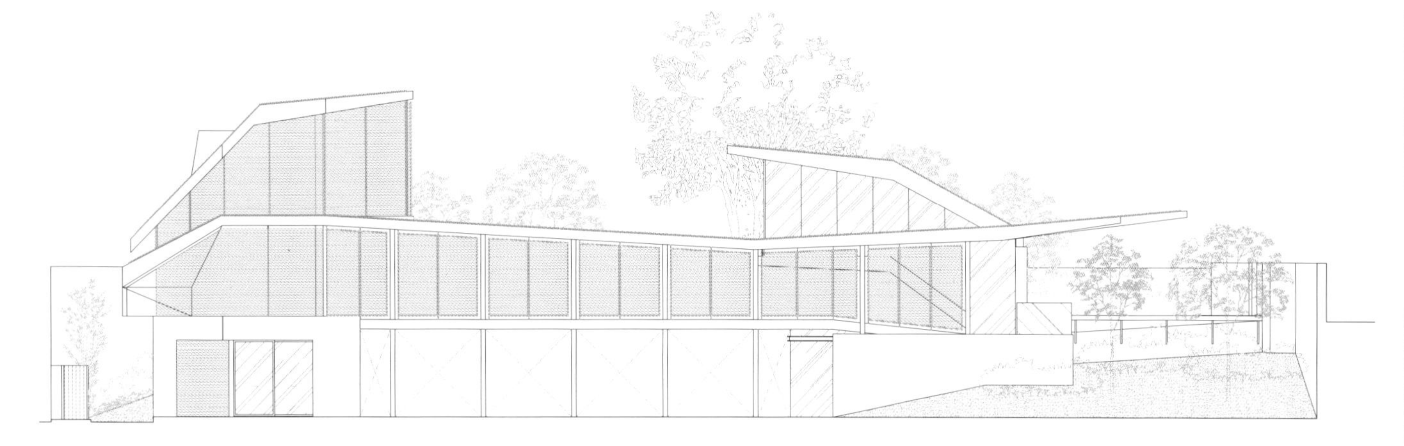

Elevation 02　立面图 2

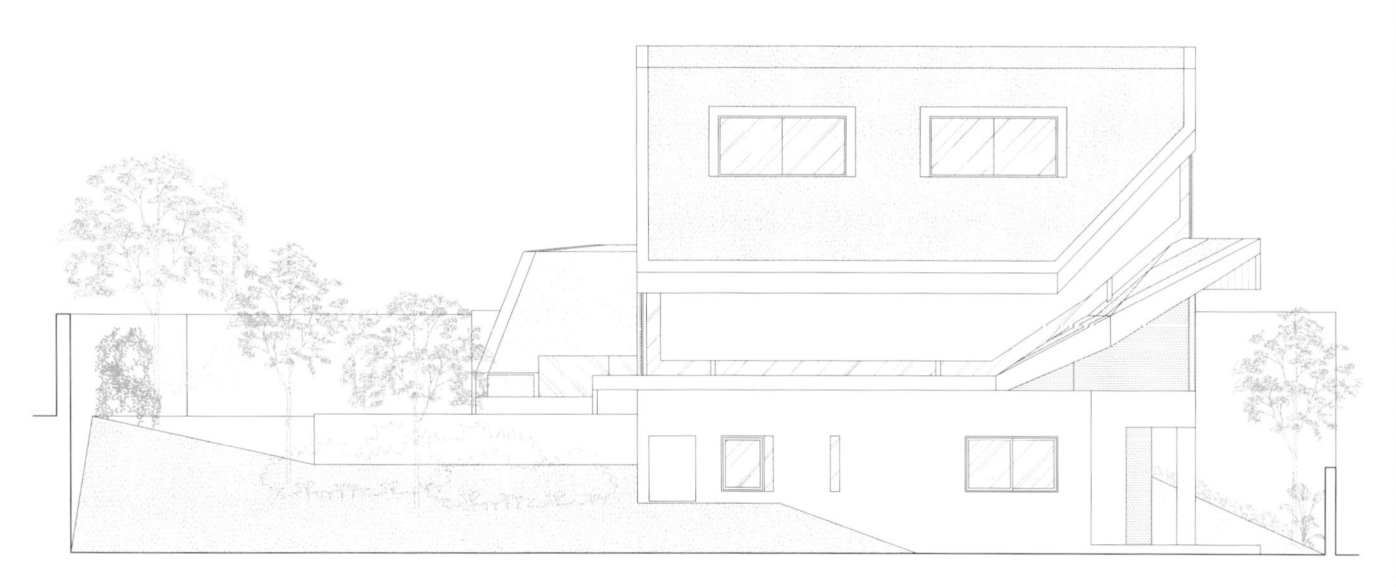

Elevation 03　立面图 3

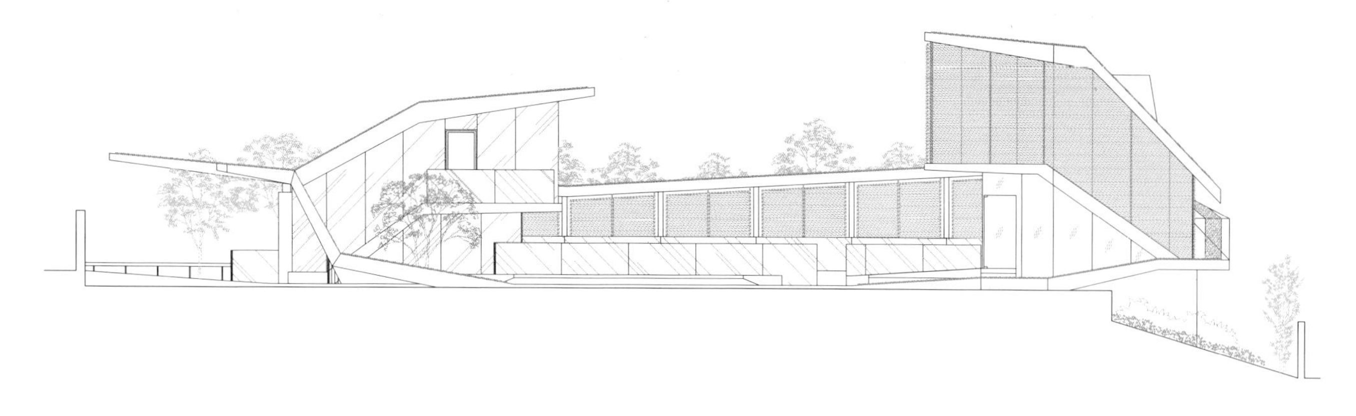

Elevation 04　立面图 4

项目概况

该别墅位于新加坡中部著名的 Caldecott 区，既拥有当地民居的轻松氛围，又享有麦里芝自然保护区的无限美景。然而，因为地块另一侧临近繁忙的高速公路，所以设计师需要解决噪声干扰的问题。

建筑设计

项目所在地拥有新加坡少见的起伏地形，建筑地块低于街道标高一层楼左右，因此这里所有住宅的入口都设在建筑的二楼。建筑师充分利用了地块的起伏进行设计，将大部分生活空间围绕下沉的庭院设计，来应对噪声干扰，同时增强私密性。建筑师将传统古朴的田园民居生活融入现代家居，起伏的屋顶草坪给人舒畅的大自然体验。泳池建在屋中央，体现了建筑师的严谨与低调，但又具有不可低估的建筑表现手法，层层递进，让偌大的别墅充满了现代生态与中央庭院的感觉。光线可以随意流淌，延续大自然美好的风光。

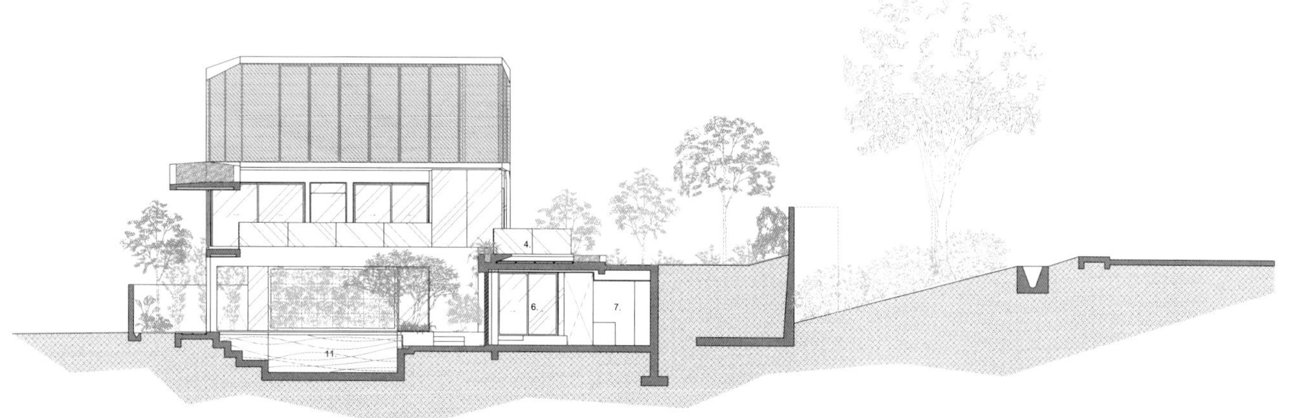

4. Deck
6. Dining Room
7. Dry Kitchen
11. Swimming Pool

4. 平台
6. 餐厅
7. 厨房干区
11. 泳池

Section AA　剖面图 A A

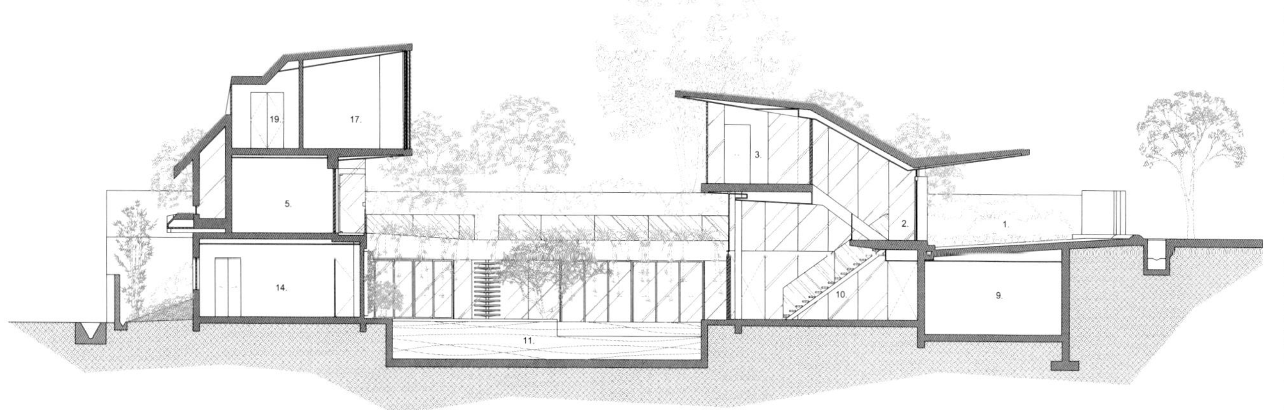

1. Car Porch
2. Entry Foyer
3. study
5. Bedroom
9. TV Room
10. Living Room
11 . Swimming Pool
14. Junior master suite
17. Master sulte
19. Wardrobe

1. 车库
2. 入口门厅
3. 书房
5. 卧室
9. 电视房
10. 客厅
11. 泳池
14. 小主人套房
17. 主套房
19. 衣橱

Section BB　剖面图 B B

Location: Singapore

Architectural Design: K2LD Architects

Floor Area: 590 m²

Photography: Patrick Bingham Hall

项目地点：新加坡

建筑设计：K2LD Architects

建筑面积：590 m²

摄影：Patrick Bingham Hall

Keywords 关键词

Twin Circular Columns 双圆形柱

Visual Transparency 视觉和谐

Reasonable Space Organization 布局合理

The Screen House

屏风住宅

Architectural Design

The house was planned as a 2 parts - a longitudinal block with a bridge connecting to an elliptical object. The strategy of the design is to merge the building with surrounding elements. Sited between the existing bungalow and a contemporary building, the screen House aims to recede from the existing bungalow and engage its neighbors and the central axis. The many trees on site in the background give an opportunity to incorporate a vertical language. A variety of timber screens and stone modules are integrated into the design of the facade. The double volume spaces inside and outside of the house are expressed by the tall and slim twin circular columns, which mimic the surrounding trees and create a visual transparency to the landscape.

建筑设计

该建筑分成两个部分，纵向的房屋主体和旁边的椭圆建筑以桥相连。其设计理念是让建筑与周围的各种元素融为一体。建筑地块位于一栋平房和一栋同期的楼房之间。宅子意在减小平房对其的影响，使其更加融入周边景观和中轴线。后院的树林与宅子形成了一种立体上的映衬，木材和石头都成为了宅子外景设计的组成部分。高细的双圆形柱很好地诠释了内外空间，柱子外形与周围的树相似，创造出了景观上的视觉和谐。

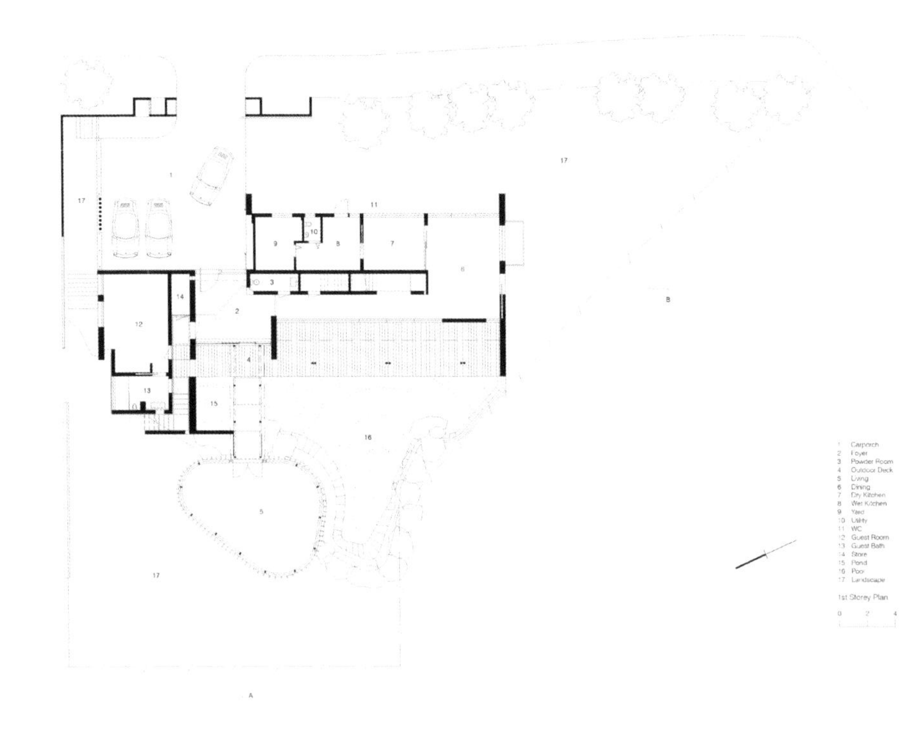

1st Floor Plan　一层平面图

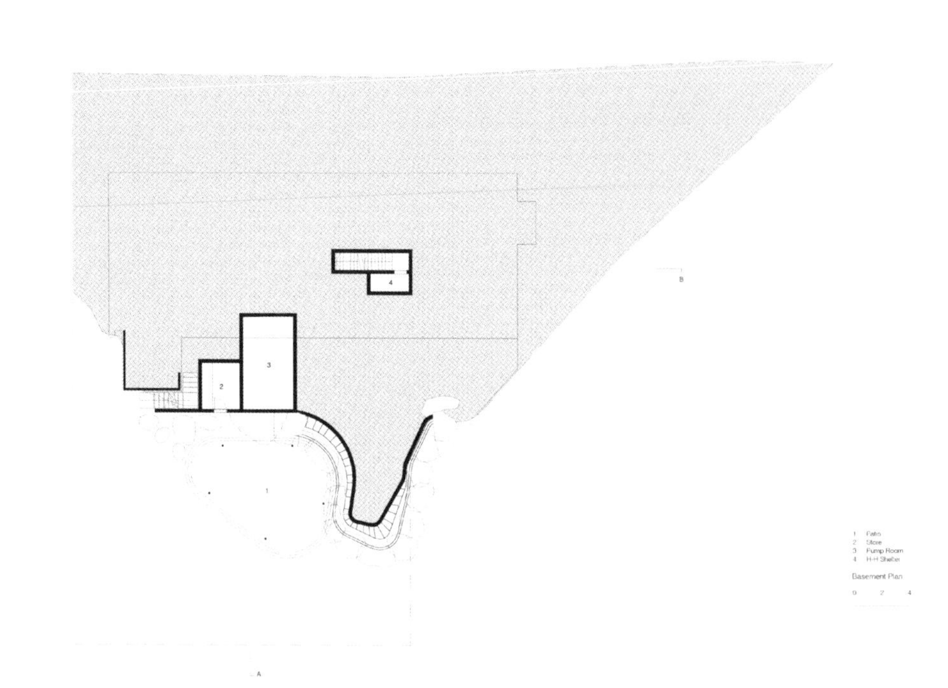

Basement Plan　地下层平面图

East Elevation　东向立面图

North Elevation　　北立面图

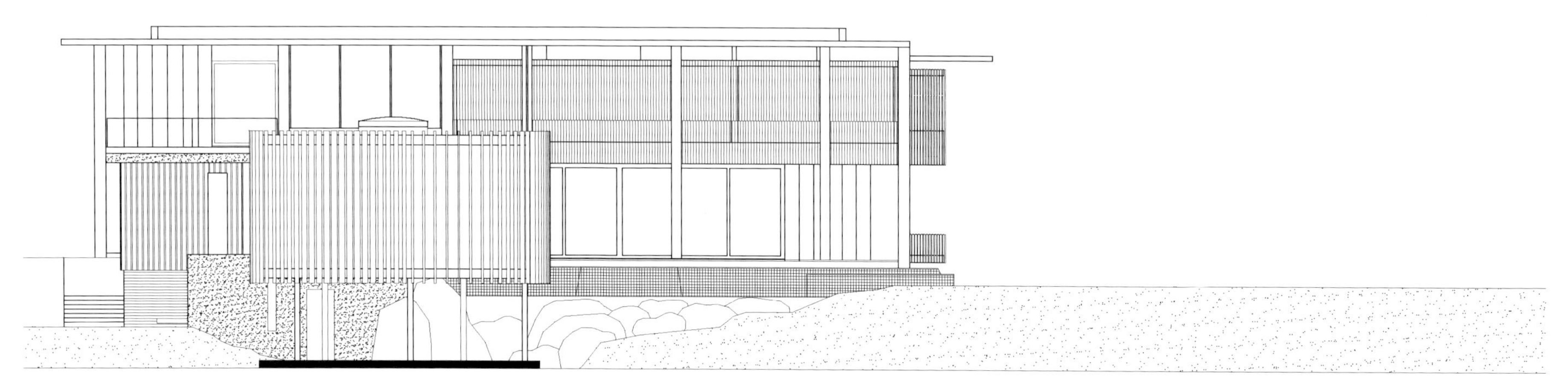

West Elevation　　西立面图

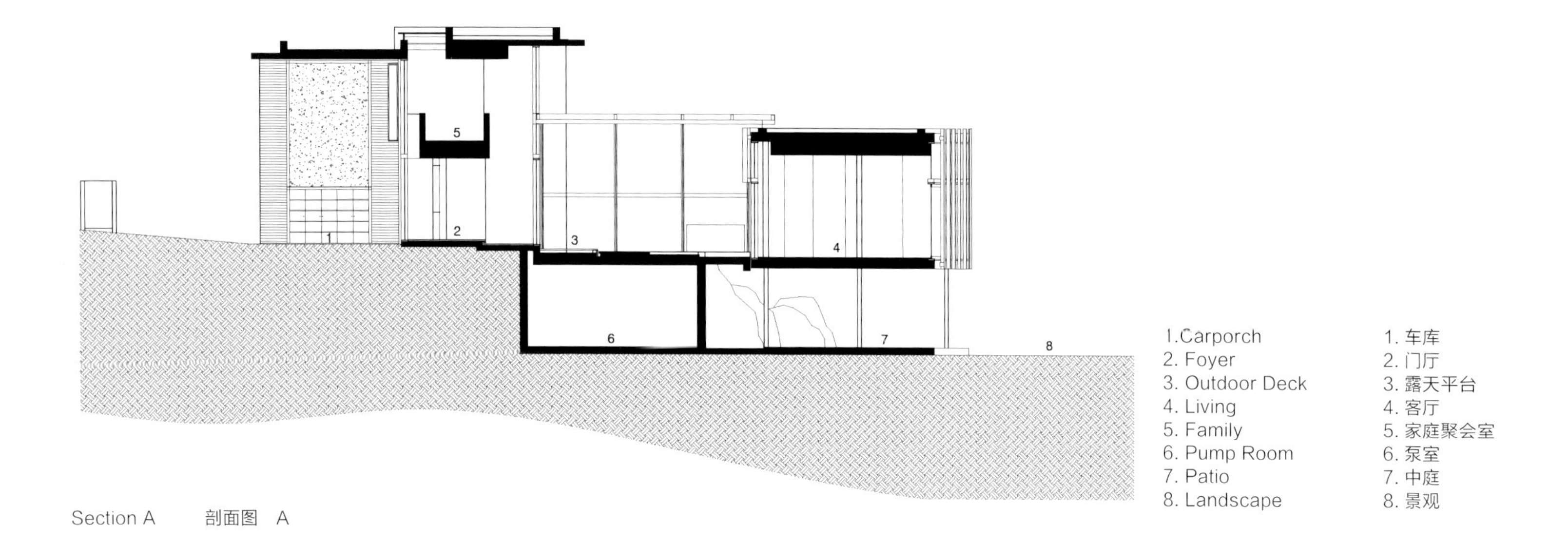

Section A　　剖面图　A

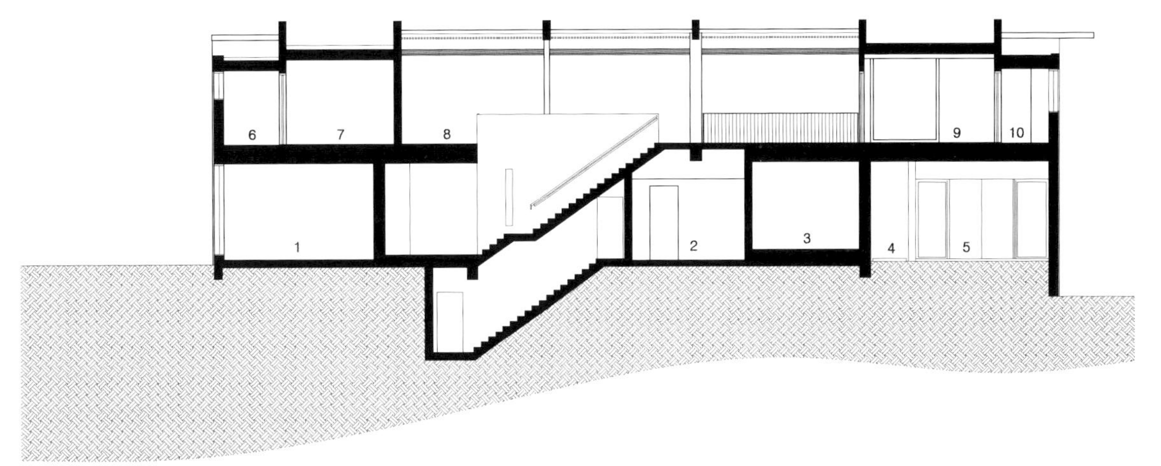

1.Dining
2. Powder Room
3. Foyer
4. Store
5. Guest Bedroom
6. Master Bath
7. Master Bedroom
8. Family
9. Bedroom
10. Bath

1. 餐厅
2. 化妆间
3. 门厅
4. 储物间
5. 客房
6. 主浴室
7. 主卧
8. 家庭聚会室
9. 卧室
10 浴室

Section B　　剖面图　B

Location: 81 Cashew Road, Singapore

Architectural Design: Red Bean Architects

Project Team Members: Teo Yee Chin, Suchada Kasemsap, Liow Zhengping, Noelia Somolinos

Construction Company: Poh Sia Construction & Engineering Pte LTD

Land Area: 478 m^2

Floor Area: 710 m^2

项目地点：新加坡 Cashew 路 81 号

建筑设计：Red Bean 建筑师事务所

设计团队：Teo Yee Chin, Suchada Kasemsap, Liow Zhengping, Noelia Somolinos

建设公司：Poh Sia 建筑工程私人有限公司

占地面积：478 m^2

建筑面积：710 m^2

Keywords 关键词

Singular Masses 体块各异

Reinforced Concrete 混凝土

Folded Slabs 折叠木板

Block House

积木石屋

Overview

This is a new bungalow for a large extended household—a couple with 3 grown up sons, each with his own family. In addition, amenities such as a karaoke room, a mahjong room, and a roof-top lap pool were requested for. The site is on the crest of a rising road and had unobstructed views of Bukit Gombak.

Architectural Design

This was a house that needed space, all 5 storeys from the basement to the attic. The resultant mass could result in a small tower, which the architects deemed as proportionally inappropriate for a house. There was thus a need to break up the verticality of the form.

The architects saw the house as 3 tectonic geological masses, pushing and straining against conformity. Frozen in a moment when they begin to slide apart from one another, the singularity of each block is expressed resulting in a horizontal articulation of the house.

The structural details required to achieve this led to an involved study in reinforced concrete. By upstand beams, structural walls and folded slabs, concrete was both the structural system as well as the material aesthetic. Square windows of varying sizes were fitted within the concrete walls. Moving through the house involved varied encounters with the surrounding greenery through these frames.

81

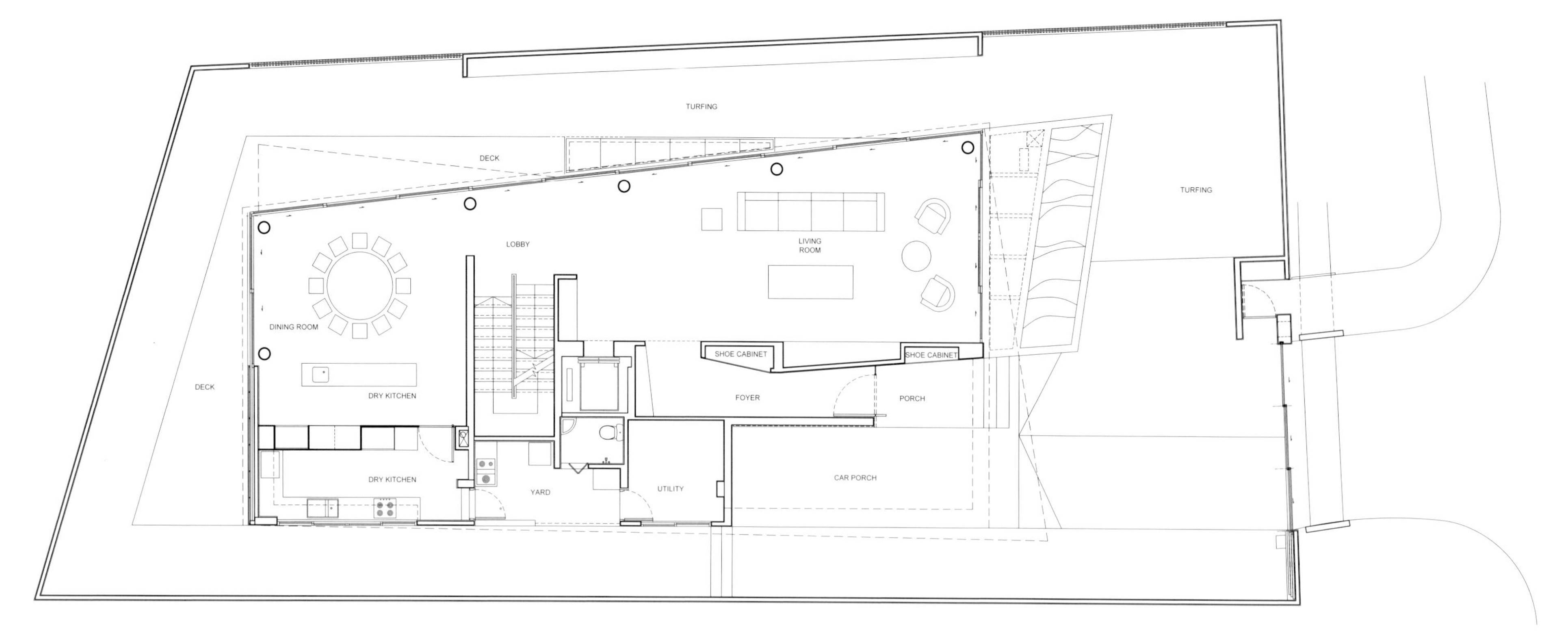

1st Storey Plan　一层平面图

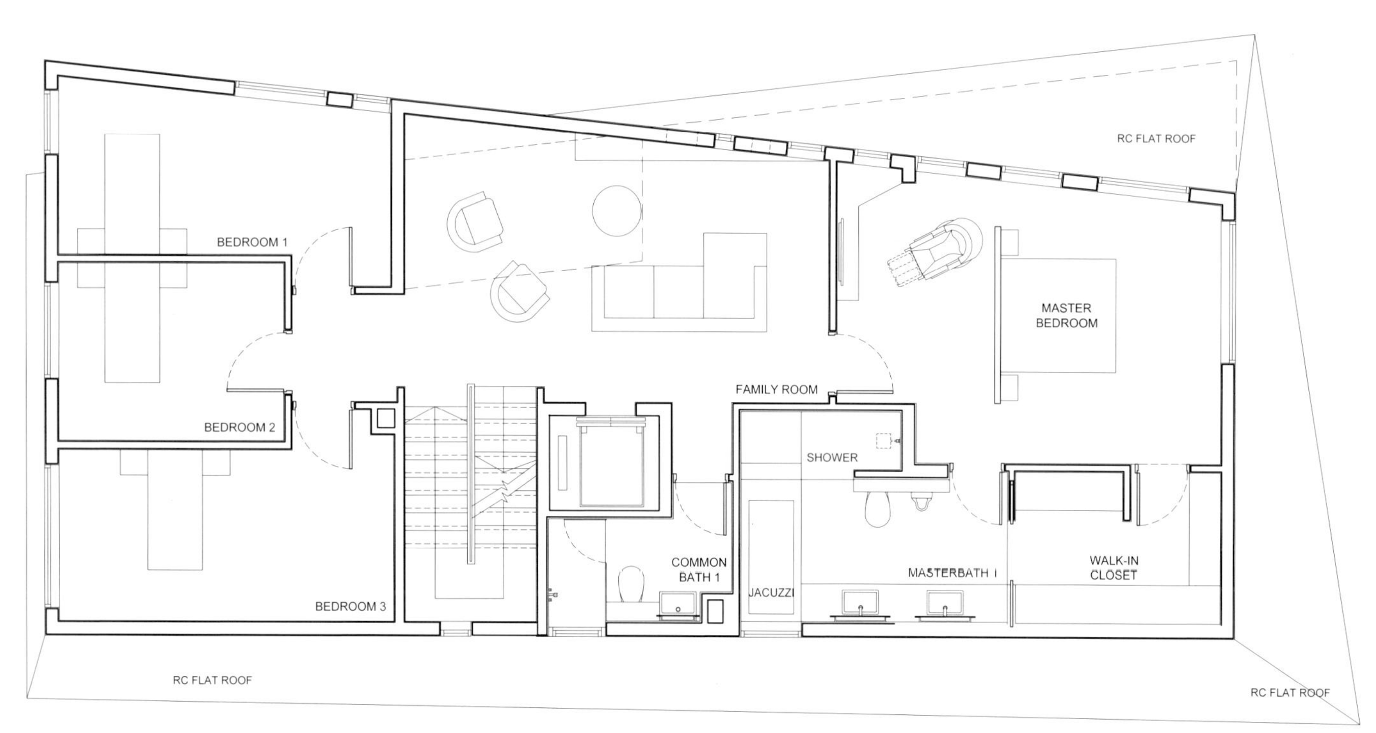

2nd Storey Plan　二层平面图

项目概况

该项目是为一个由父母以及三个各自成家的儿子组成的大家庭设计的住宅。客户还要求要有卡拉 OK 房、麻将房、屋顶泳池。该项目位于这段上坡路的至高点，尽享甘柏山美景。

建筑设计

该住宅分为地下一层，地上四层，客户还希望能拥有更多空间。房子原有的结构会使得它成为一个狭窄的塔楼，而建筑师们认为这样的比例完全不合理。于是，建筑师需要打破原有房子的垂直线条。

建筑师们将房子锯成三个形状各异的地质结构的体块，定格在仿佛将它们彼此抽离出来的那一瞬间。三个体块的奇特也使得整个房子看起来是被水平地连接起来似的。

为了能完美地呈现出房子的细节，就需要仔细衡量钢筋水泥的比重与结构，包括直立梁、承重墙以及折叠木板。混凝土不仅成为了房子承重的一部分，也担当了房子外部的装饰材料。设置在混凝土墙之间的各种大小的方形窗户，使住户在房子的各处都能欣赏窗外的绿色景致。

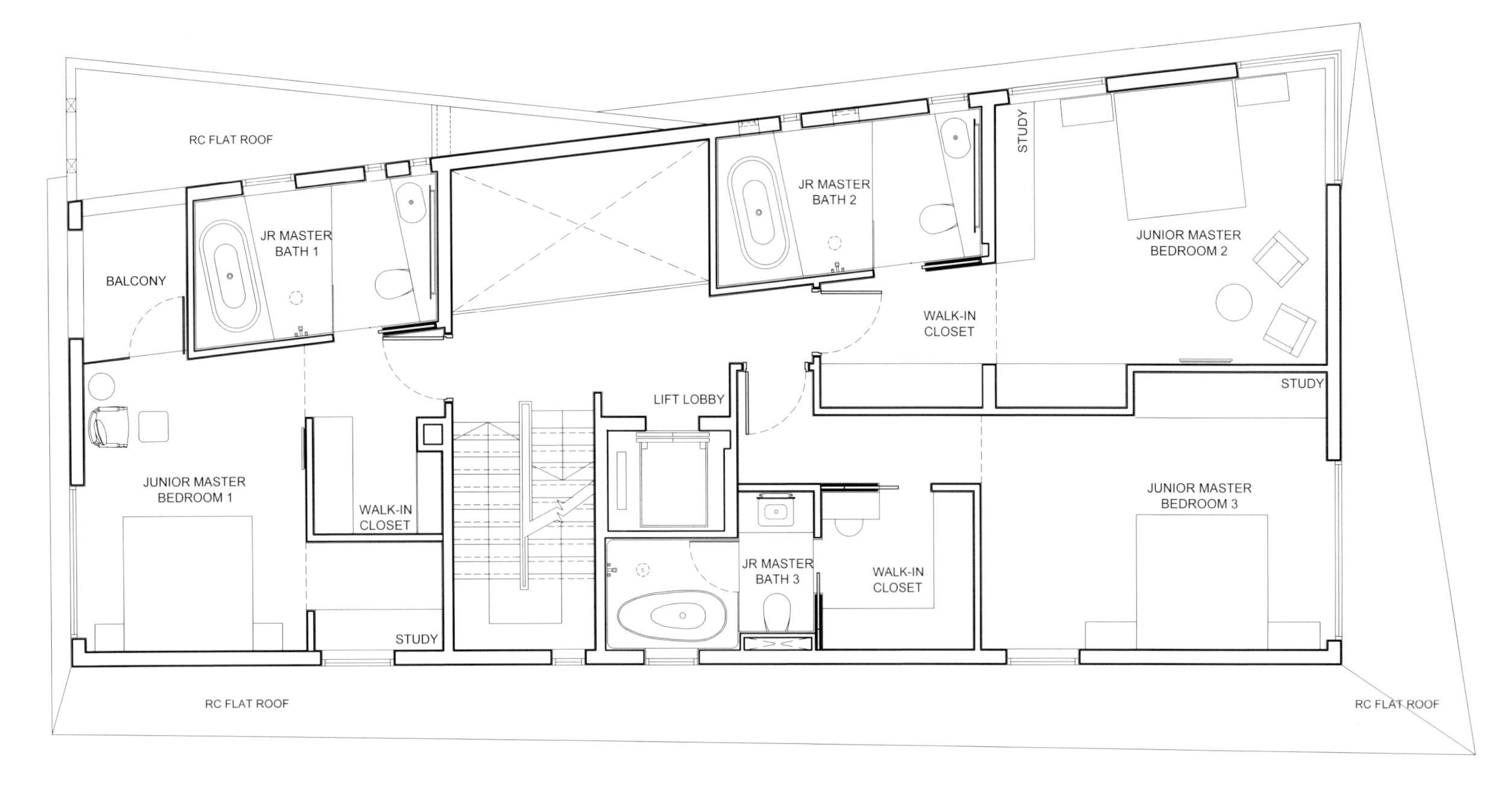

3rd Storey Plan　三层平面图

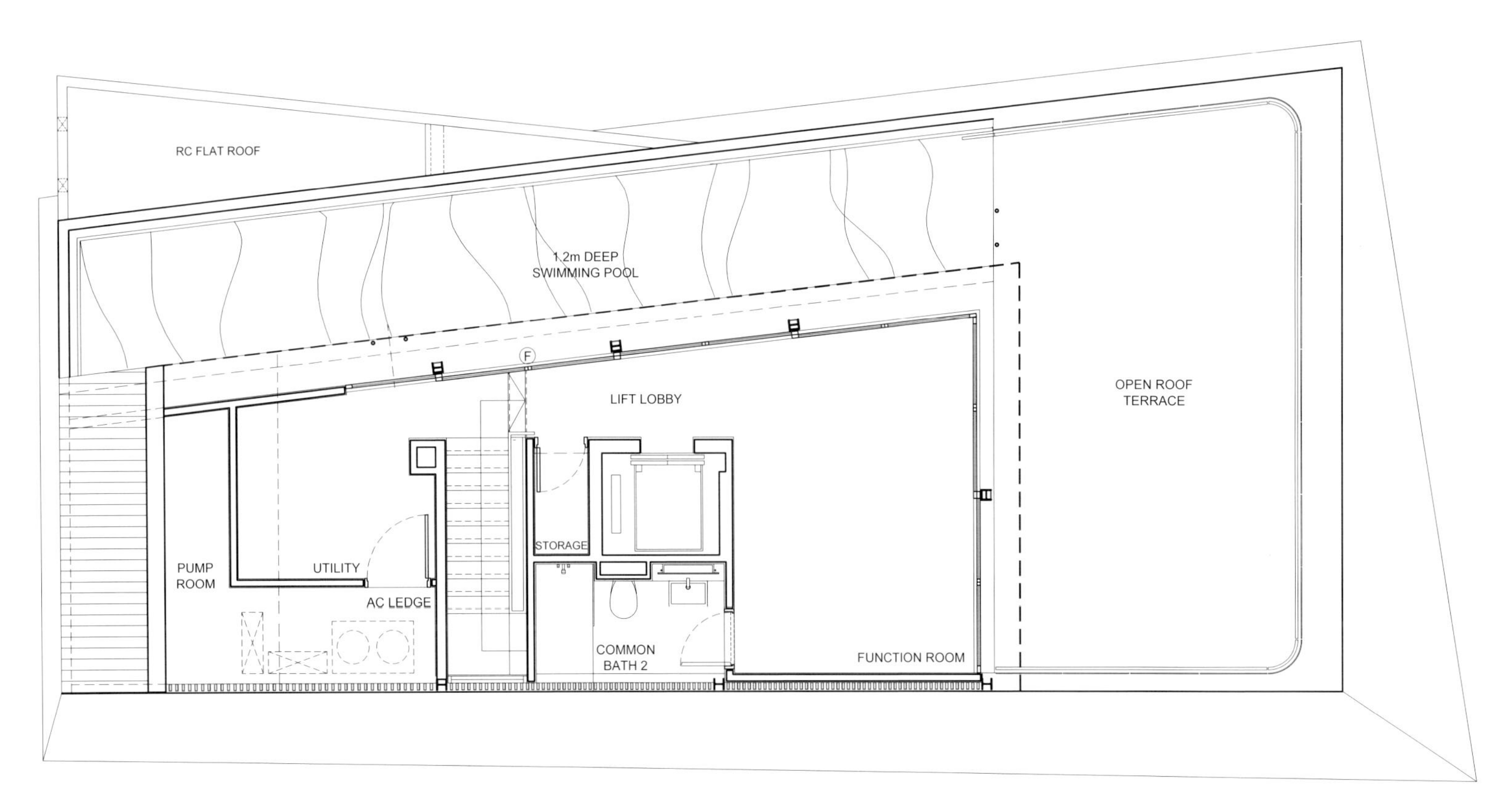

Attic Storey　阁楼层平面图

Street Elevation 街道立面图

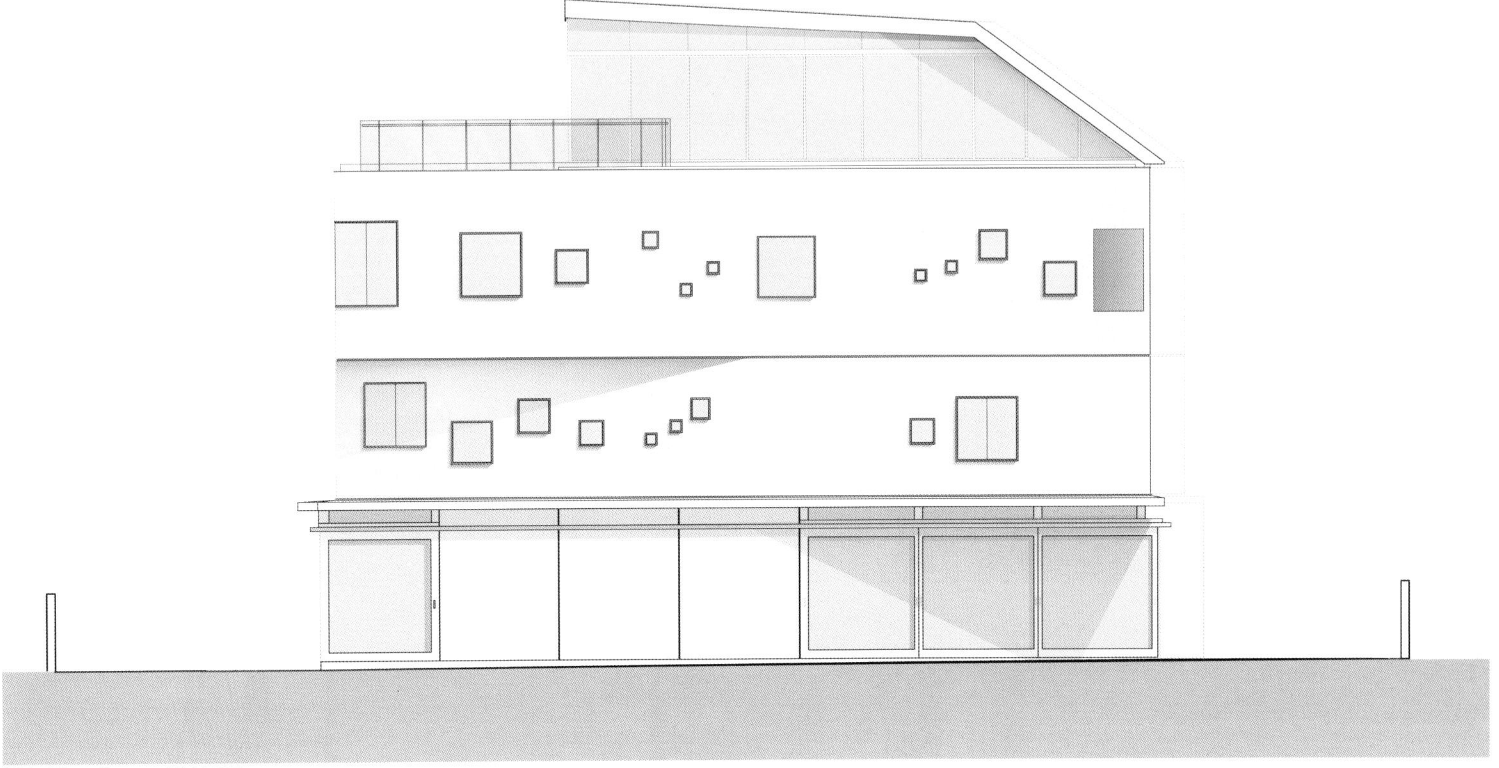

Side Elevation 侧立面图

Location: Holland Park, Singapore

Architecture/Landscape/Interior Design: Tierra Design (S) Pte Ltd

Photography: Patrick Bingham-Hall, Amir Sultan

项目地点：新加坡荷兰园

建筑 / 景观 / 室内设计：Tierra 设计事务所

摄影：Patrick Bingham-Hall, Amir Sultan

Keywords 关键词

Site Plan 组合布局

White Boxes “白色盒子”

Shared Garden 公共庭院

8 Box House

盒式住宅

Overview

Seated at the base of a gentle hill in Holland Park, Singapore, 8 Box House is one of six distinctive bungalows of the Lien Villas Collective. To achieve variety and color, each bungalow on the property was assigned to a different architect. Tierra Design was commissioned to develop the site plan and create what is now the iconic 8 Box House.

Site Plan and Architectural Design

Today, the 8 Box House holds its own amongst its four "sibling" houses, and the conserved patriarch house. The bungalow is a cluster of eight giant white boxes, each containing a room of a different size and directional orientation.

The final massing offers a unique view of the house from different angles. It features internal spaces that are interconnected physically and visually to each other, as well as to the surrounding views outside. The 8 Box House project presented a unique opportunity for Tierra Design to advance the value and appreciation of distinctive architecture in Singapore.

View Corridor

Central to the final site plan is a primary view corridor that pivots from the entrance of the patriarch bungalow. While it provides a welcome stretch of shared garden spaces for the various families to enjoy, the concept of the verdant spine also unlocked tremendous new value for the site. The plan allowed for the parcelling of the component plots in such a way that a sixth bungalow could be added to the site.

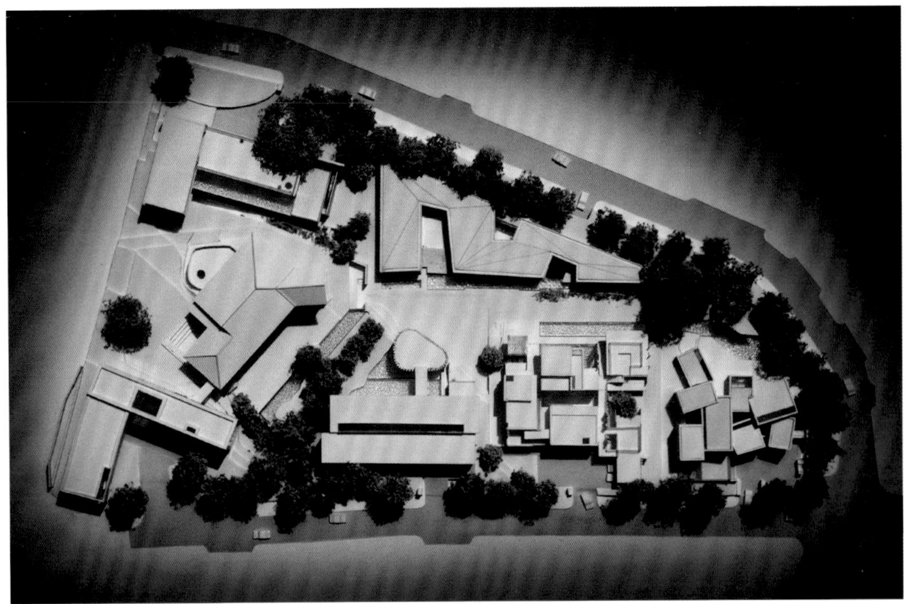

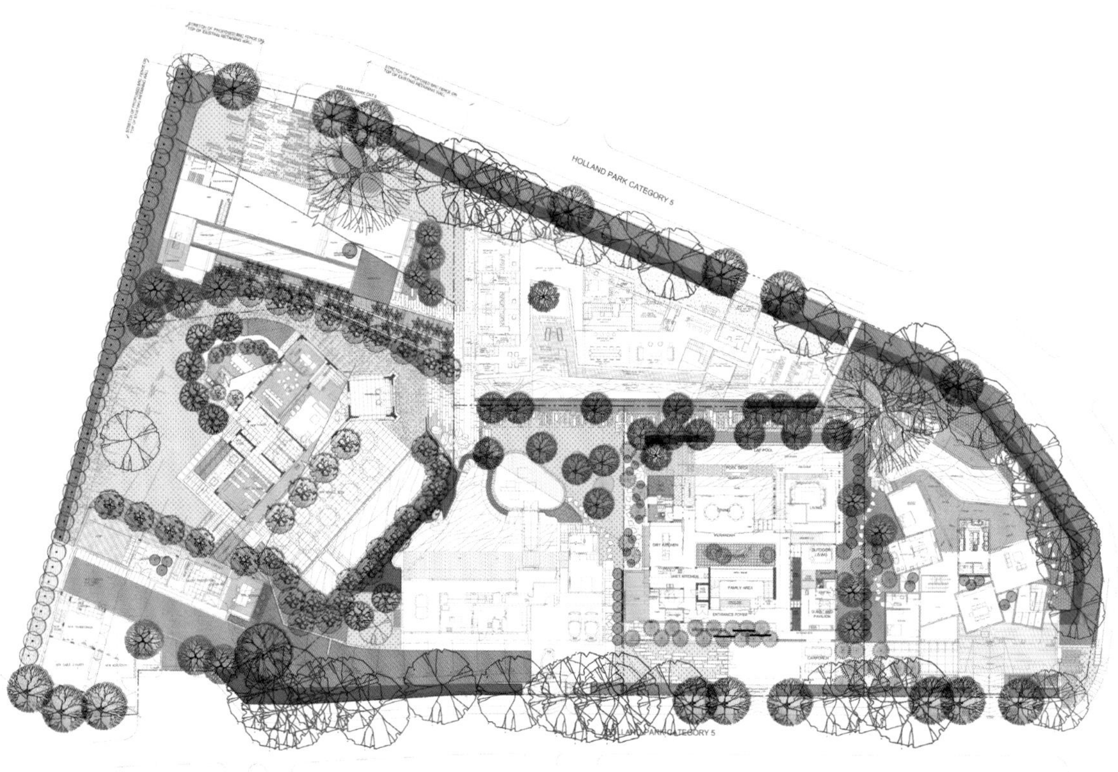

项目概况

项目位于新加坡荷兰园一座缓坡上，是 Lien 别墅系列之一。为了达到风格多样、色彩丰富的效果，这六个地块分别委托不同的设计师进行设计。Tierra 设计事务所承接了该项目的总体规划和这座标志性的盒式住宅设计。

组合布局与建筑设计

如今这座盒式住宅置身于其他四座姊妹住宅和一座族长府邸的包围中，看上去如同由八个巨大的白色盒子堆砌而成。每个“盒子”大小不一、朝向各异，提供不同的生活空间。

这种组合使住宅呈现出不同的视角。建筑内部在空间上和视觉上各个相连，同时与周围环境也保持联系。该住宅设计对 Tierra 设计事务所来说是个很好的实践机会，使之为新加坡特色建筑添上了浓墨重彩的一笔。

景观走廊

根据总体规划，场地中部将建一座景观走廊，连接族长府邸入口，并通往公共庭院。这种“绿色脊梁”的设计理念极大地提升了场地的价值。通过合理规划，六座特色别墅既相互独立，又呼应彼此，形成一个完整的系列。

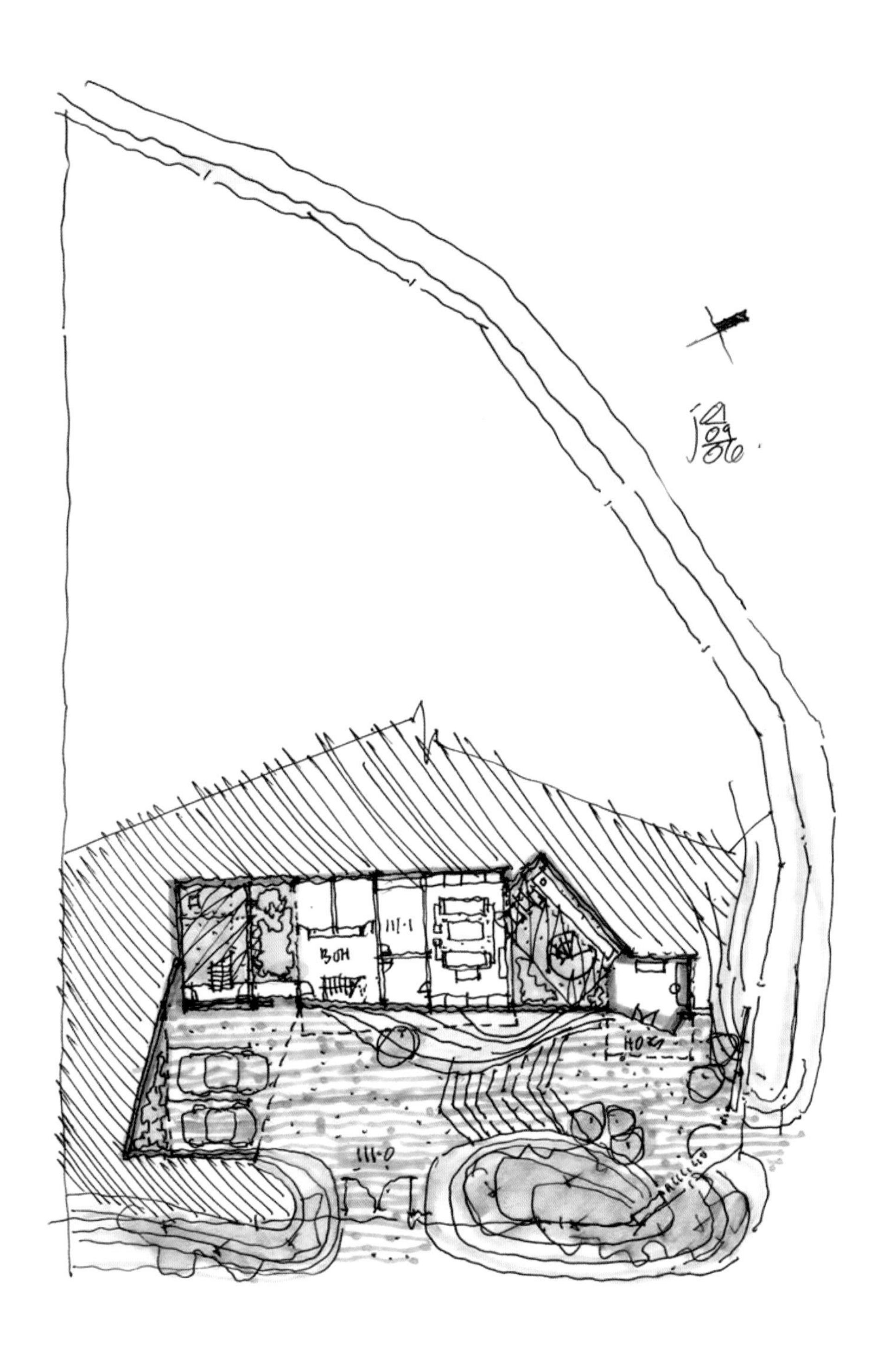

1. Driveway
2. Main entrance
3.Ante 1
4.Carporch
5.Entrance
6.Ante 2
7.Wet kitchen
8.Breakfast
9.Guestroom
10. Guestbath
11.Household shelter
12.Laundry & store
13.Store
14.Maid's room
15.Maid's bath
16.Landscape courtyard
17. Service yard

1. 车道
2. 主入口
3. 走廊 1
4. 车库
5. 入口
6. 走廊 2
7. 厨房湿区
8. 早餐区
9. 客房
10. 客房浴室
11. 遮蔽区
12. 洗衣房和储物间
13. 储物间
14. 佣人房
15. 佣人浴室
16. 景观庭院
17. 杂作坊

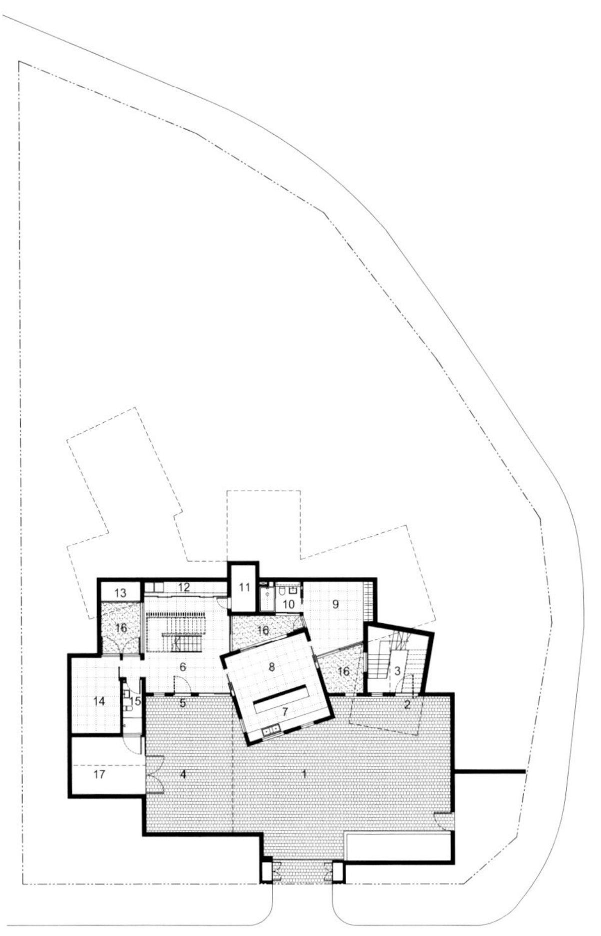

Basement Plan　地下层平面图

1.Foyer1
2.Living room
3.Dining room
4.Dry kitchen
5.Powder
6.Store
7.Foyer2
8.Bedroom 1
9.Bath 1
10. Walk-in-closet
11.Master bedroom
12.Master bath
13.Pool deck
14.Swimming pool
15.Patio
16.Void above courtyard
17.Garden

1. 门厅 1
2. 客厅
3. 餐厅
4. 厨房干区
5. 化妆间
6. 储物间
7. 门厅 2
8. 卧室 1
9. 浴室 1
10. 衣帽间
11. 主卧
12. 主卧浴室
13. 泳池甲板
14. 泳池
15. 中庭
16. 庭院入口
17. 花园

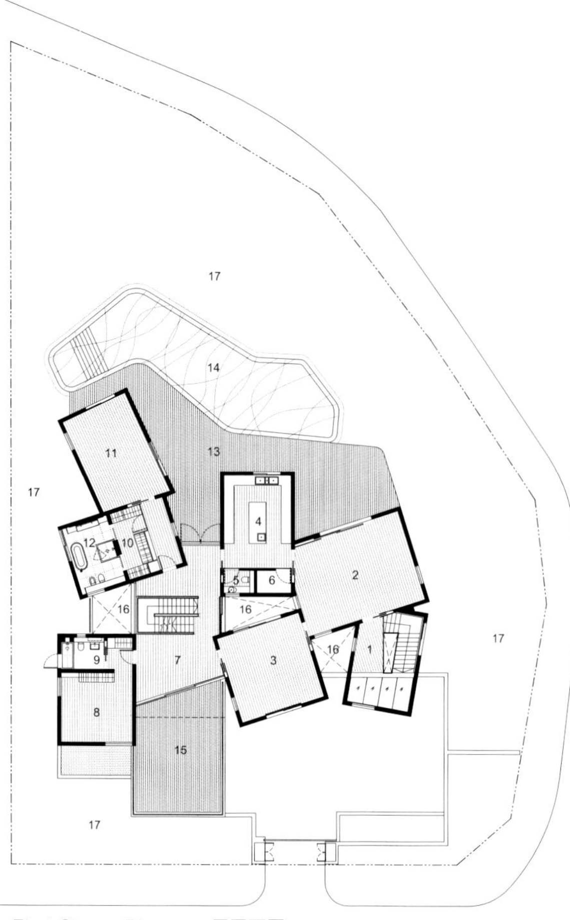

First Storey Plan　一层平面图

1.Family
2.Bedroom 2
3.Bath 2
4.Bedroom 3
5.Bath 3
6.Void over courtyard
7.Metal roof with skylight
8.Metal roof

1. 家庭间
2. 卧室 2
3. 浴室 2
4. 卧室 3
5. 浴室 3
6. 庭院入口
7. 带天窗的金属屋顶
8. 金属屋顶

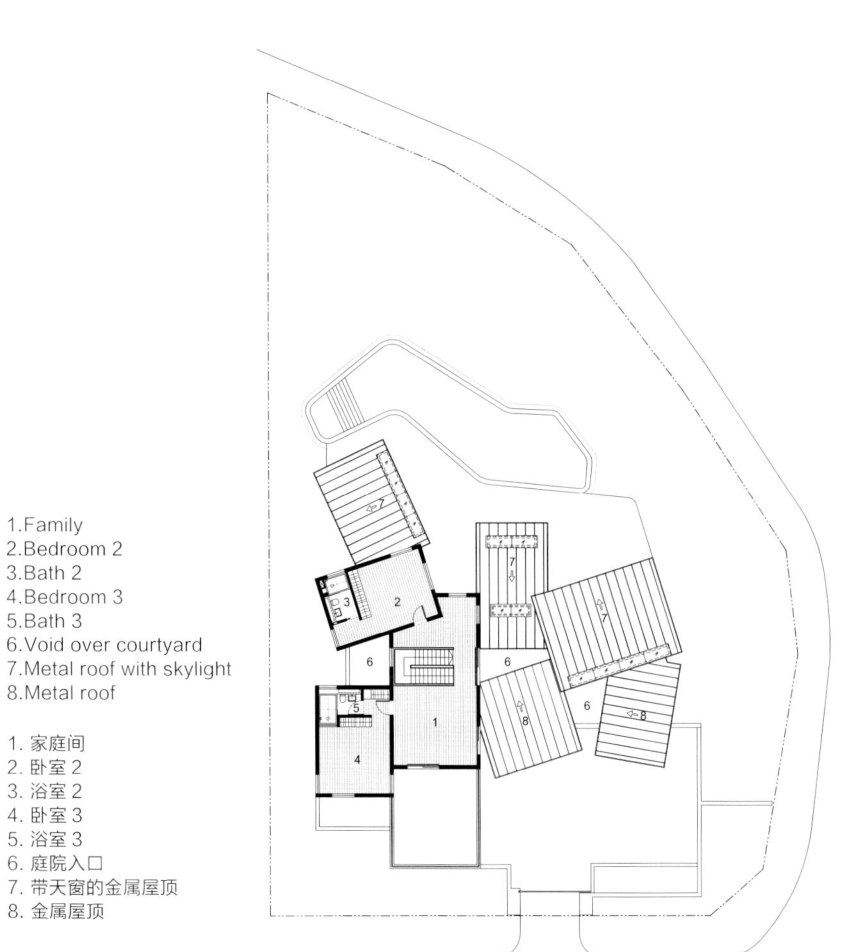

Second Storey Plan　二层平面图

1 Metal roof with skylight
2 Metal roof

1. 金属屋顶天窗
2. 金属屋顶

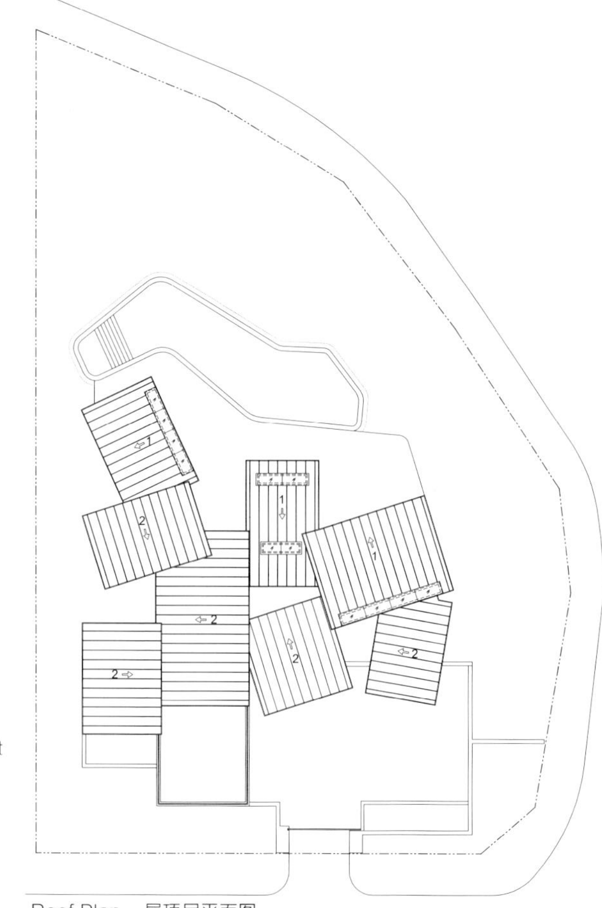

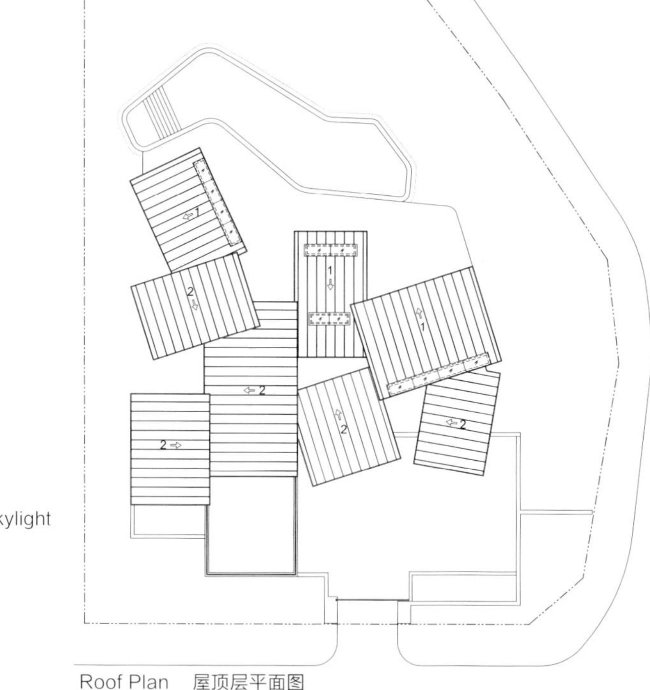

Roof Plan　屋顶层平面图

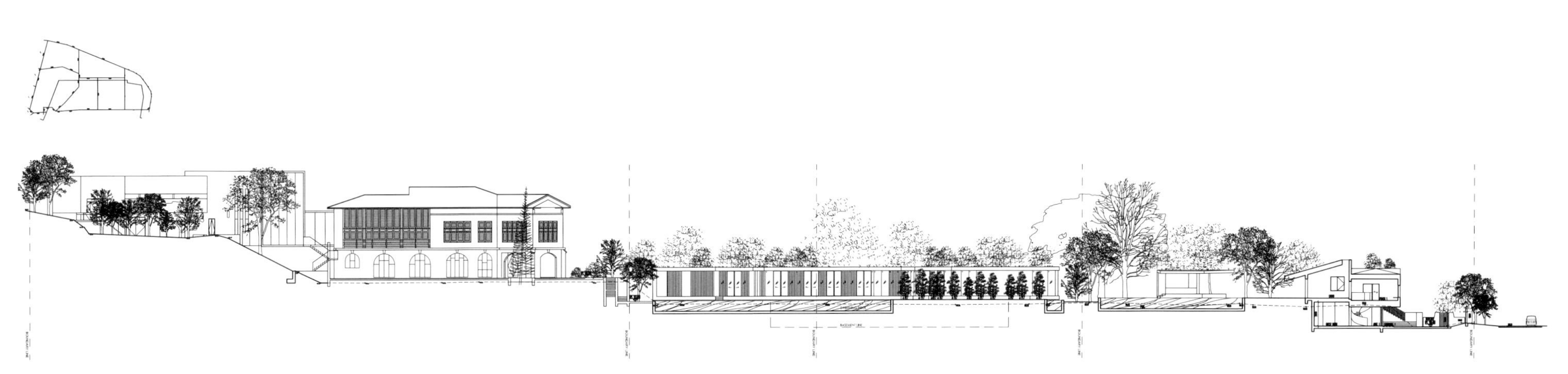

Section 4　剖面图 4

Section 5　剖面图 5

Front Elevation　正立面图

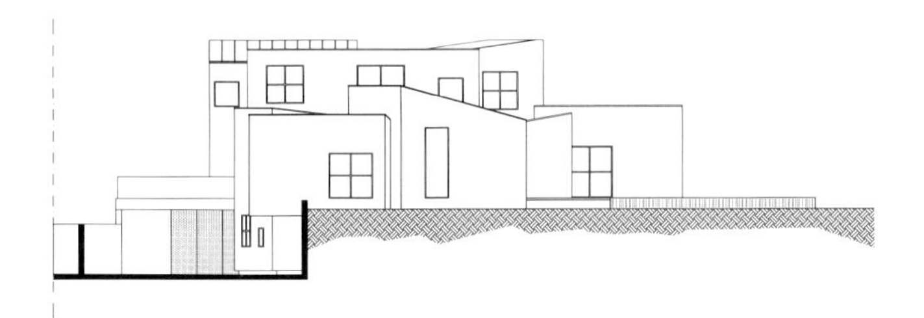

Right Elevation　右立面图

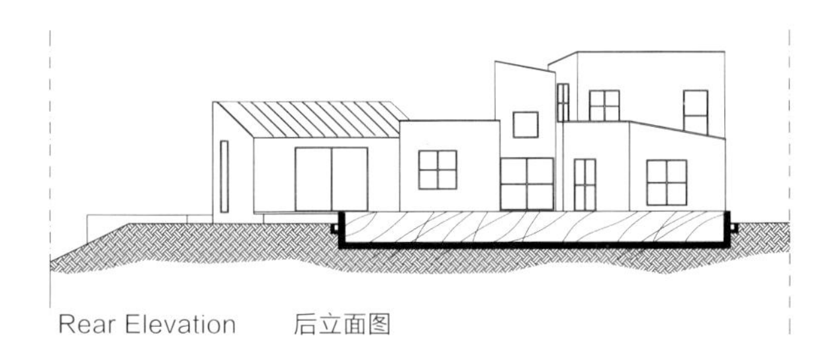

Rear Elevation　后立面图

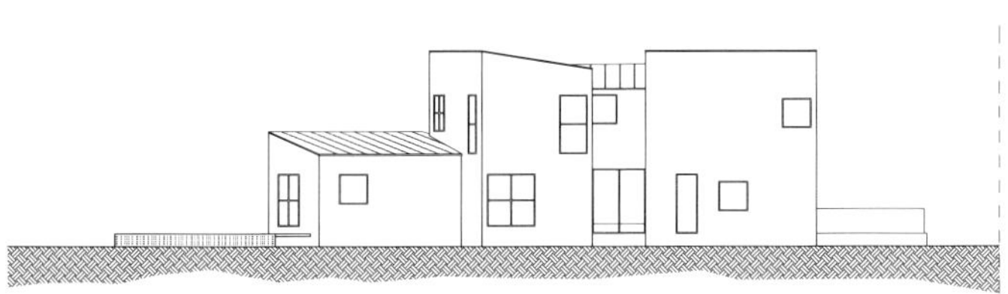

Left Elevation　左立面图

Beach Houses

滨海别墅

Sea View　Natural Ventilation　Multiple Functions

海景景观　自然通风　功能多样

Location: Sentosa Cove, Singapore

Architectural Design: Wallflower Architecture + Design

Design Team: Robin Tan + Cecil Chee + Sean Zheng

Photography: Albert Lim

项目地点：新加坡圣淘沙湾

建筑设计：新加坡桂竹香建筑 + 设计

设计团队：Robin Tan + Cecil Chee + Sean Zheng

摄影：Albert Lim

Keywords 关键词

Rectangular Project 矩形建筑

Entrance Facade 入口立面

Enclosing Wall 围护墙

Sun Cap House

太阳帽住宅

Overview

Although the site was devoid of any development during the inception of the project, it was foreseen that the future built environment would be dense with neighbouring residences barely metres away on either side. The tropical sunlight falling on this resort island could also be harsh and intense but the proximity to the sea also blesses it with breezes that tend to channel through the waterways that are unique to the cove. Most properties along the waterway which also affords them best view, and the narrow rectangular project site was no different.

Architectural Design

In response to the projected urban density and the site's local environment, the home is designed with a thick, nine-metre high wall that forms the entrance facade which wraps around to continue along the sides. Like the pulling back of a curtain to reveal the view, the walls terminate as they approach the waterway where thereafter an inner enclosing structure of paneled glass continues, projecting toward a pool and garden. The massive, enveloping entrance and side walls are essentially a thermal and privacy filter. The wall occludes views from inquisitive neighbours but encourage the passage of breezes that find their way through the house rather than around it by deliberate vertical slotting dividing the enclosing wall into free-standing segments. The slotting also helps to filter natural light into the house and soften the impact of the harsh sunlight. The secondary glass paneled enclosure within but set away from the enveloping walls is designed to slide away so that the impression of width does not terminate at the glass line but are extended to the tall side walls. The impression of space however goes even further, for the slots in the walls reveal landscaping that extends beyond. The walls are parallel to but do not meet the eaves of the roof; one-metre wide gap invites sunlight to wash down onto planting and greenery that thrive on either side of the wall blurring the distinction of an "inside" "outside" demarcation. Though a vertical surface, the rough plastered texturing of the wall catches light streaming in from the gap above and diffuses it into the living spaces.

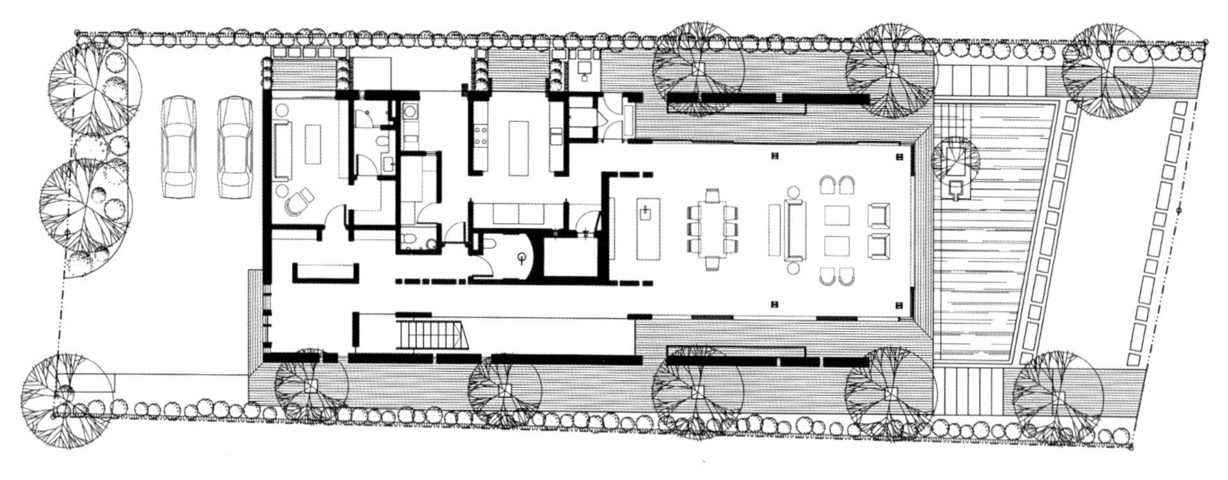

First Storey Plan　一层平面图

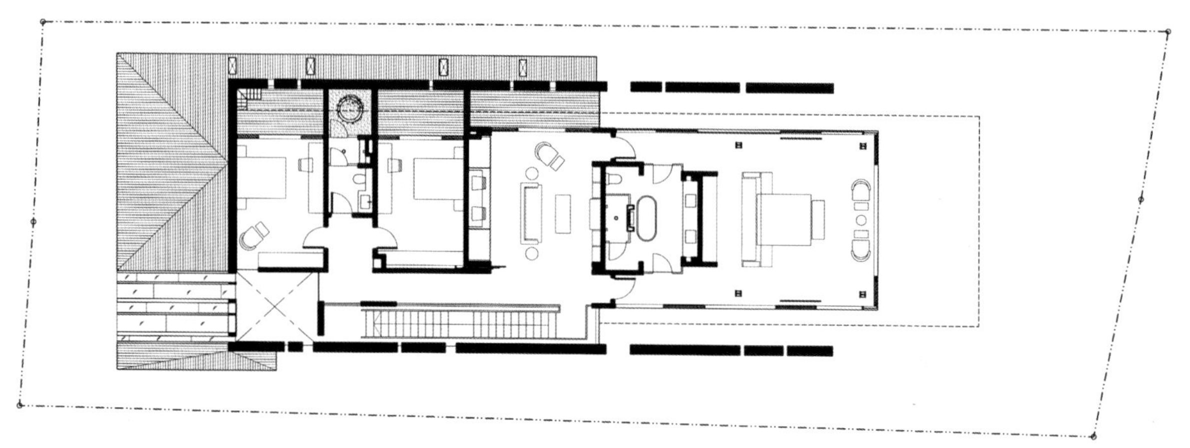

Second Storey Plan　二层平面图

项目概况

尽管该别墅在项目初期缺少发展，但可以预见，未来的建筑周围的住宅将十分密集，两侧的距离仅仅几米之遥。洒落在这个度假岛屿的热带阳光，也有可能是恶劣且强烈的，但是临近大海可享受海风，海风可通过海湾独一无二的水道吹拂而来。沿着水道而建的房屋通常也有最好的视野，这座狭长的矩形建筑也不例外。

建筑设计

为了适应项目所在地的城市密度和当地的环境，这座别墅设计了一面 9 m 高的厚墙，形成入口立面，从四周绕道两侧。就像拉开窗帘显示窗外的视野一样，厚墙在接近水道时终止，水道后面一个镶嵌玻璃的内围护结构延续着，朝着池塘和花园悬出。厚重的包围着的入口和侧墙是隔热和保护隐私的关键。侧墙挡住了邻居好奇的视线，但是可供海风直接穿过房屋，而不是通过刻意垂直开槽把围护墙分成独立的段数。开槽也有助于过滤自然光线进入室内，缓解阳光刺眼的影响。第二层玻璃板在围护墙内，但是远离围护墙设计成滑动的形式，所以对宽度的印象不是停止在了玻璃液面线，而是延伸到高高的侧墙。空间感甚至更进一步，墙上开槽可以看到延伸出去的景观。墙壁是平行的，但是跟屋顶的边缘没有接触。1 m 宽的鸿沟使得阳光倾泻而下照射植物和绿化带，墙两侧的植物蓬勃生长，模糊了室内外界限的明显区别。虽然是垂直表面，墙壁粗糙的灰泥纹理聚集从上端的缺口流淌进来的光，继而扩散到各个生活空间。

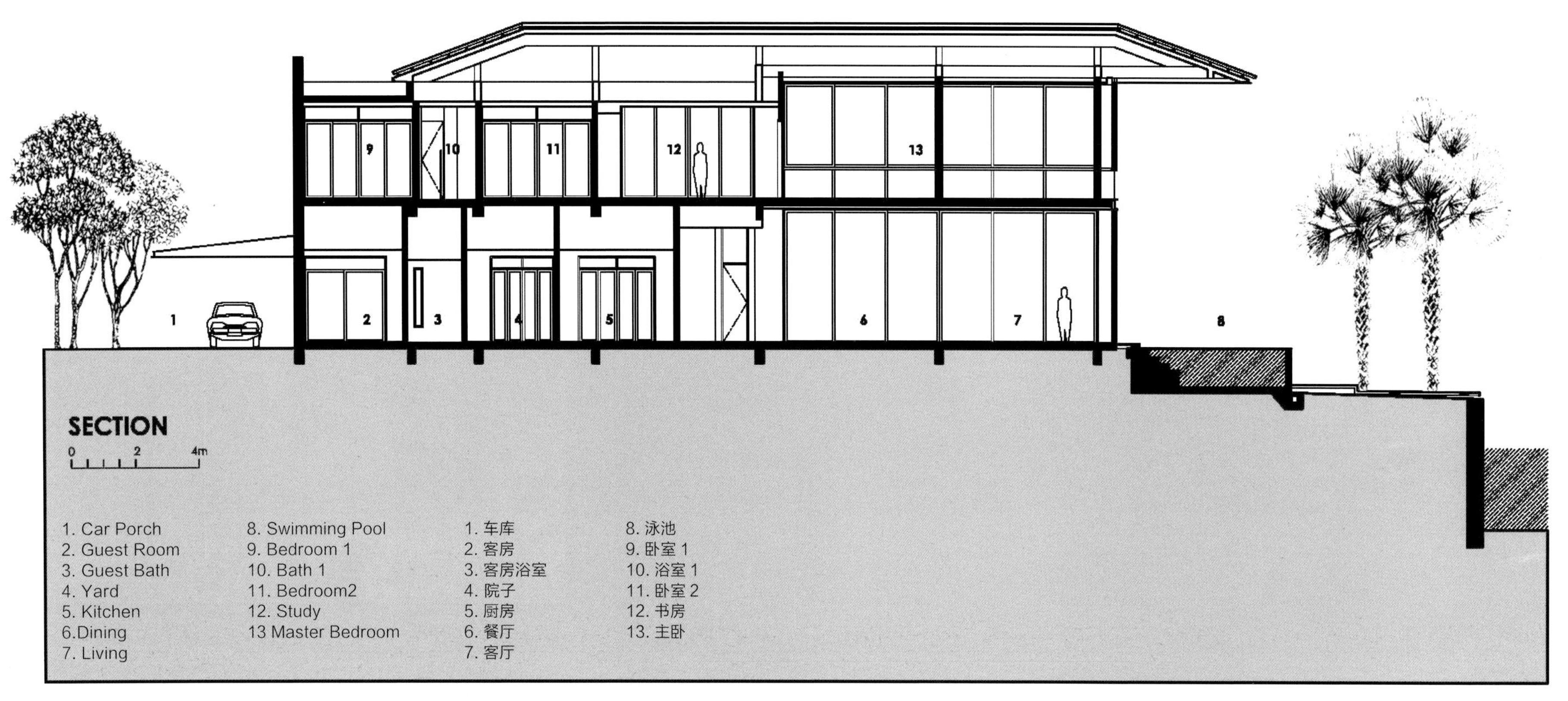

Section 剖面图

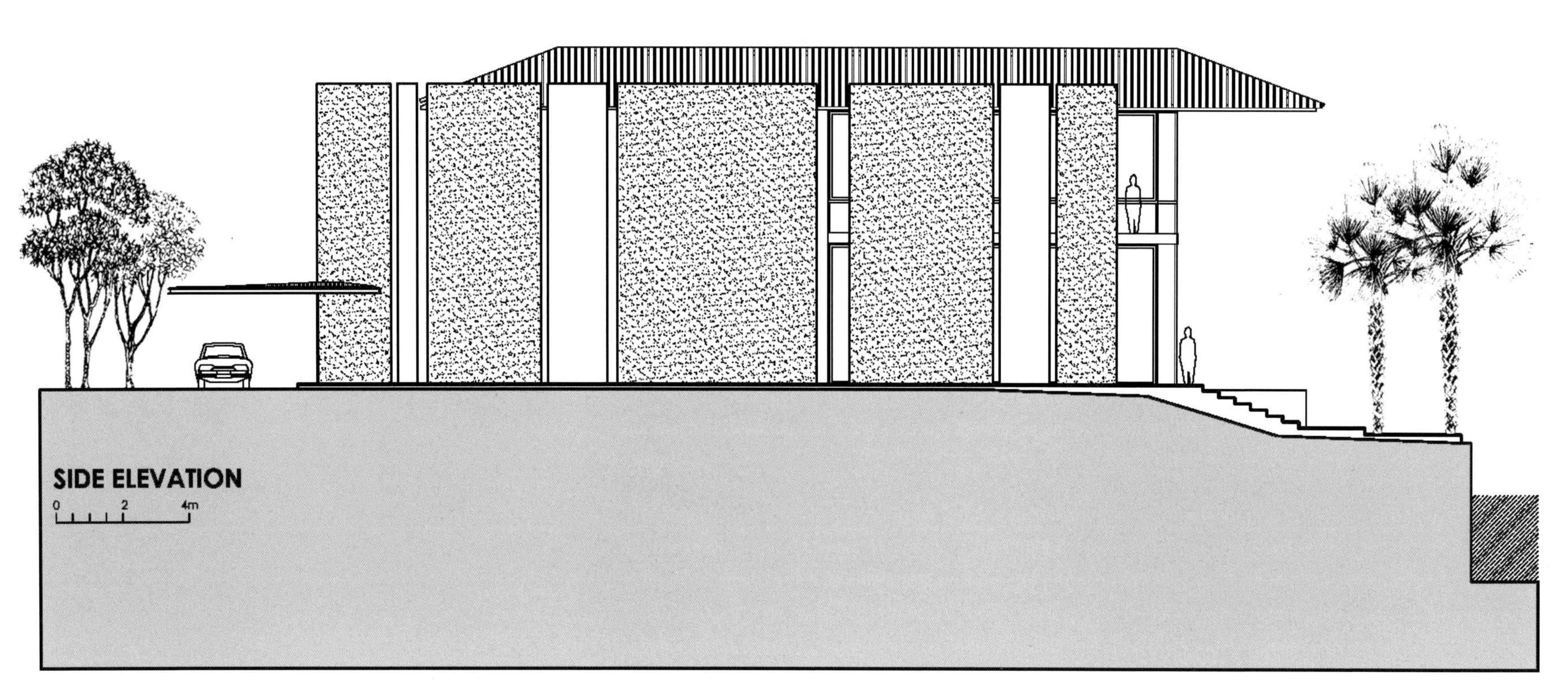

Side Elevation 立面图

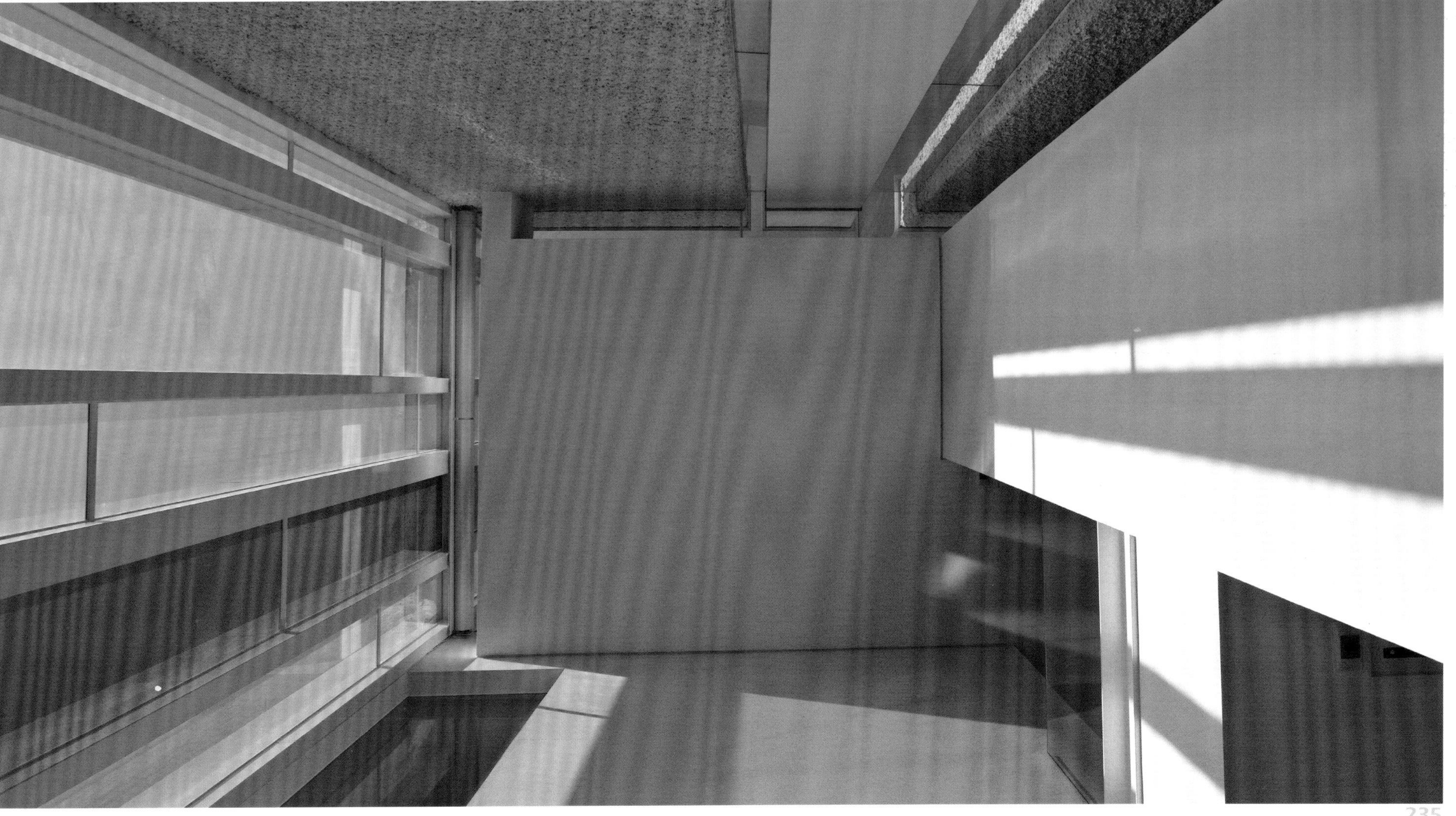

Location: Singapore

Architectural Design: a_collective

项目地点：新加坡

建筑设计：a_collective

Keywords 关键词

L-shaped L形布局

Open Space 开放空间

Natural Ventilation 自然通风

Sunset Terrace House

日落阳台住宅

Overview

The site for this corner bungalow house lies at a busy road junction with a long view axis towards the main road. An L-shaped building layout plan creates a private garden at the rear corner of the site. The building mass cradles this inner sanctuary, with views from the living and kitchen spaces on the ground storey centered towards the green landscaping.

Spatial Experience Pavilion

The combined living and dining room has the spatial experience of an open pavilion, with the natural elements of water on one side and green landscaping on the other. A 3.6m high teak timber ceiling allows maximum light to enter the living spaces and excellent natural ventilation when the glass sliding doors on each side are opened. As a result, the house is infused with light and air, allowing for a pleasant living experience in Singapore's tropical climate. Rain is kept at bay with a large overhanging aluminium roof which also provides ample shade in the hot afternoons.

A fully-floating feature staircase ties the two stories together in a double height volume, lined with 8m high off-form concrete walls. Borneo Ironwood timber is utilised on the second storey building facade to insulate the bedrooms as well as to break down the building mass. At the owner's requests, the timber strips are untreated, creating a personal, alluring quality to the house.

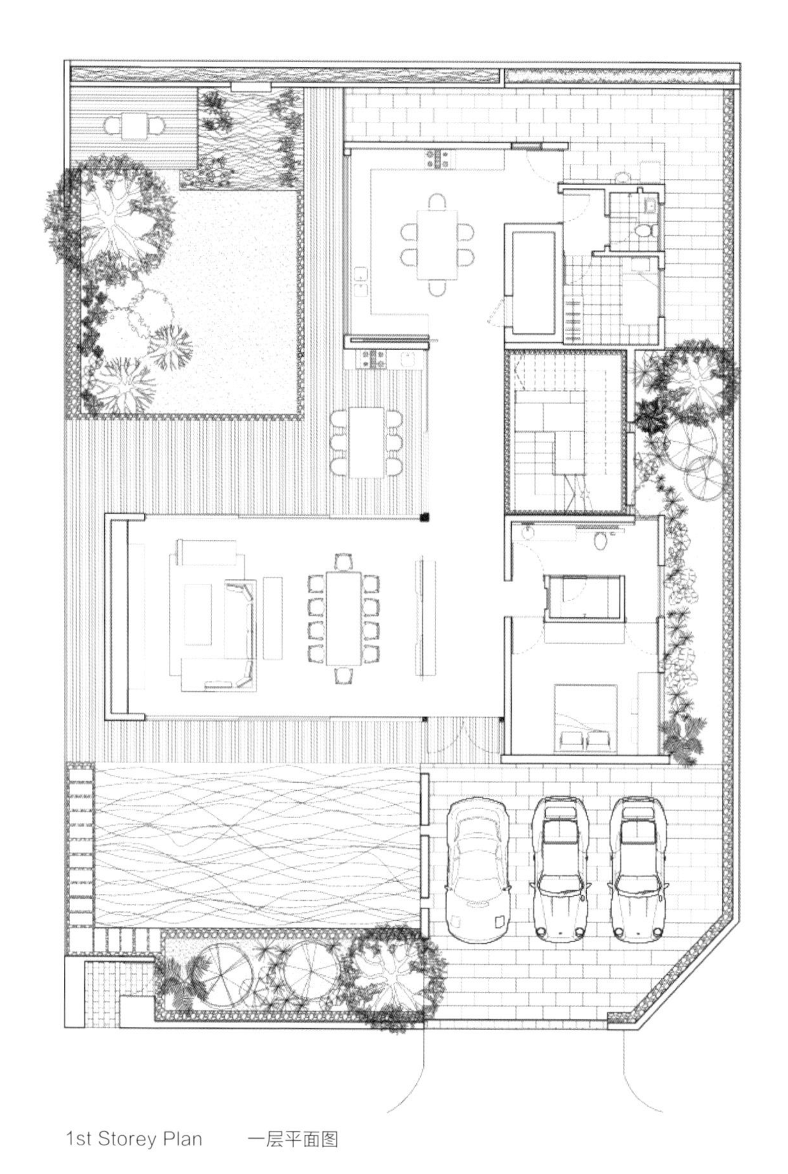

1st Storey Plan　一层平面图

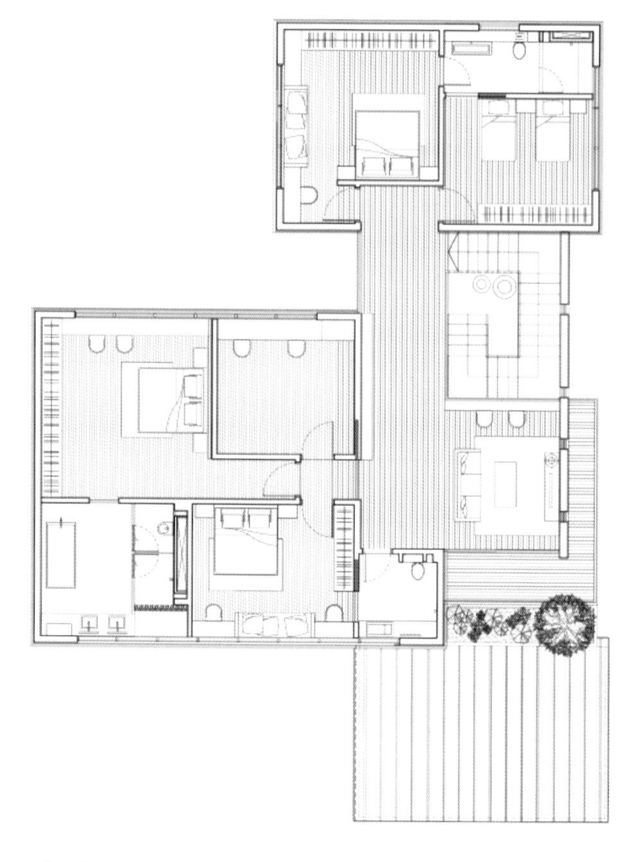

2nd Storey Plan　二层平面图

项目概况

这个平房住宅坐落在一个繁忙的主要道路的路口，拥有一个狭长的视野。L 形的建筑布局在场地后侧的角落里创建了一个私人花园。该建筑群使人内心宁静、放松，从一楼的客厅和厨房都能欣赏到绿色景观。

空间体验馆

连接在一起的客厅和餐厅形成了一个开放性的空间体验馆，一边是自然水域，另一边是绿色景观。高达 3.6 m 的柚木天花板，能保证最多的光线照进生活空间，当所有的滑动玻璃门都打开的时候也能拥有良好的自然通风。因此，房屋内部充满了自然光和空气，使住户在属于热带气候的新加坡也能有一个舒适的生活体验。大型悬垂铝制屋顶能使雨水不流进房子，在炎热的午后，也提供了一片阴凉。

在这个 8 m 高的混凝土墙支撑的双层空间里，一个类似浮动着的楼梯连接着一楼和二楼。设计师们利用婆罗木材在二楼的建筑外墙来隔离卧室，打破了建筑体量。在客户的要求下，条状的木材未经过任何处理，以创造一个私密的、诱人的住宅。

Keywords 关键词

Tube Mass 管状建筑

Roof-pitch 坡面屋顶

Natural Light 自然光线

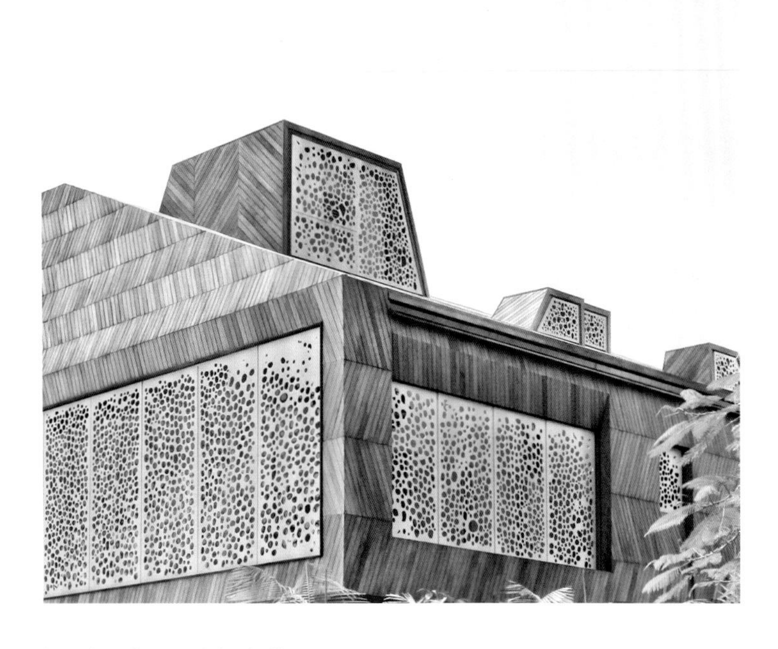

Location: Sentosa Island, Singapore

Architectural Design: Pencil Office

Lead Design Consultant: Erik G. L' Heureux AIA, LEED AP BD+C

Floor Area: 708.8 m^2

项目地点：新加坡圣陶沙岛

建筑设计：Pencil Office 事务所

设计主创：Erik G. L' Heureux AIA, LEED AP BD+C

建筑面积：708.8 m^2

Stereoscopic House

立体住宅

Overview

Located on the flat reclaimed landscape of Sentosa Island in the Singapore Straits, the Stereoscopic House is sandwiched between a natural ocean view, an artificial golf course view and two neighboring units two meters left and right. A dramatic combination of overhanging volume and structure reconfigure the relationship between environment, landscape, and view.

Design Concept

From the view of architects, the combination of a weekend resort house and a sustainable agenda drives the design strategies of the Stereoscopic House. A series of highly effective passive techniques reconcile the client's requests for size, luxury, and generous amenities while mitigating the tremendous environmental impacts of being located at 1°15': the tropical equator.

Architectural Design

A distorted tube containing bedrooms on the upper floors isolate three sisters from the adjacent neighbors, while framing dramatic views to ocean and golf course in a stereoscopic relationship. Roof-pitch codes deform the tube; creating a formal approach of deep angular overhangs and striking verandas, reducing insolation on the exterior courtyard, terrace, and living spaces.

The angular roof-pitch helps to further frame a picturesque view to neighboring islands on the third storey terrace. In response to the tropical climate of Singapore, an additional layer of timber cladding is added to the roof, accommodating angular dimensions as well as minimizing heat transfer through the kalzip roof and into interior spaces below. A western street facade alludes to the symbolic profile of a "house" in response to the demands of the client, while a dramatic eastern facade is calibrated precisely for view from two bedrooms and their shared verandah and terrace.

Diffused and reflected sunlight brightens interior spaces in the house through the use of screens attached to angular skylights sited on the roof and windows subtracted from the facades. Together, these additions and subtractions of volumes not only facilitate natural day lighting, but also create a phenomenological experience of light and shadow, as well as an architectural language, unique to the home.

107

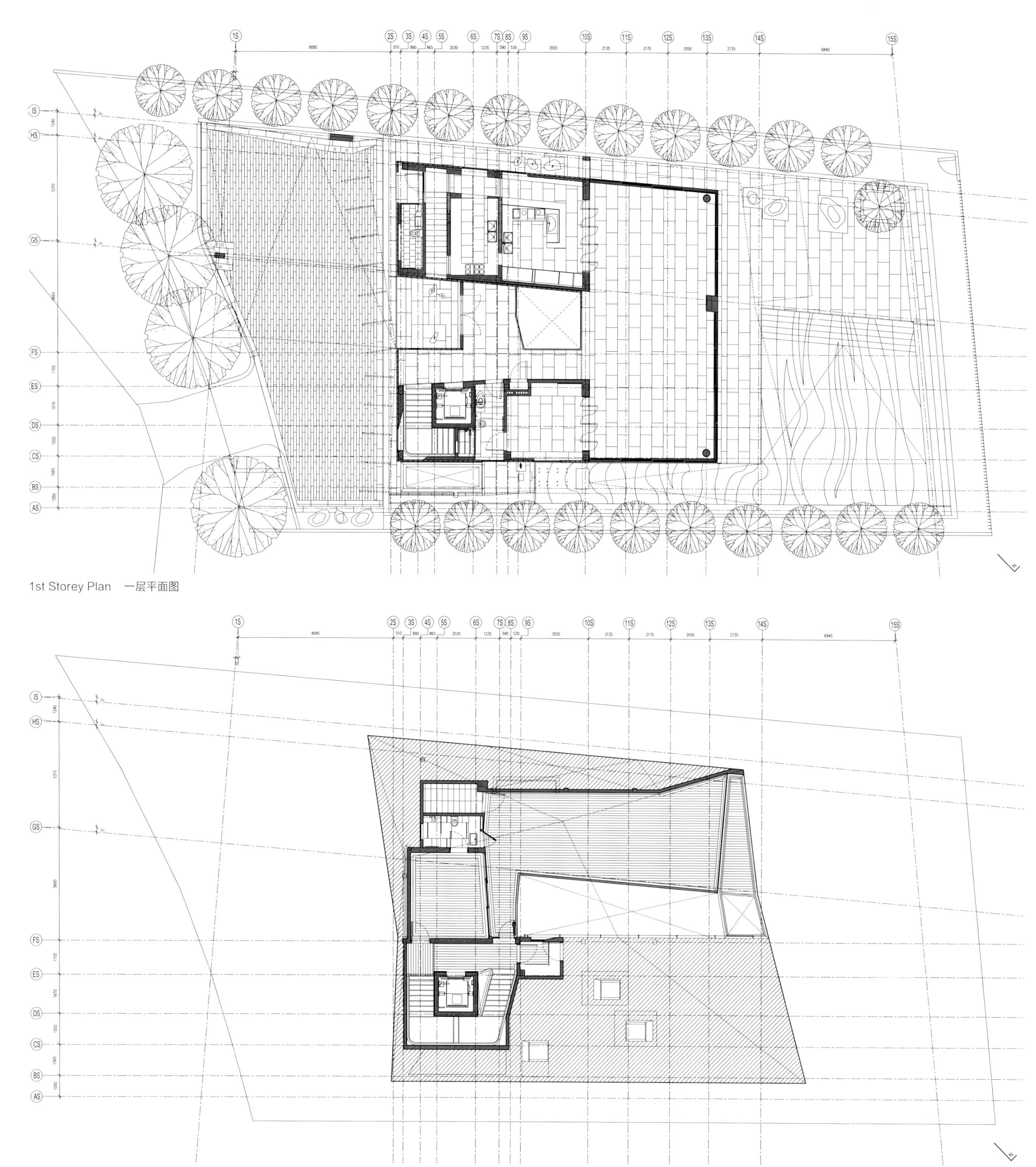

1st Storey Plan　一层平面图

Attic Plan　阁楼层平面图

项目理念

项目坐落于新加坡圣陶沙岛的平坦景观带上，这处立体住宅就如同三明治一般，处在自然海景视廊、人工高尔夫球场视廊以及左右距离两米的相邻住宅之间。悬挂的体量、改良的结构与周边环境、景观和视廊极佳地融合在一起。

设计理念

在建筑师看来，这座立体住宅的设计构思来源于将度假住宅和可持续的概念融合在一起。通过一系列的高效巧妙的技术将业主关于大小、奢华以及舒适度的要求协调处理，同时也缓和了该项目所处的赤道地区环境对建筑的巨大影响。

建筑设计

在二楼的倾斜的管状结构内是卧室，这将三个姐妹的起居生活同毗邻的住户分隔开来，同时还营造了极好的眺望海景和高尔夫球场的立体视角。屋顶的坡度给管状体赋予了新的意义，创造了一个正式的有棱角的阁楼和令人惊讶的平台。同时，减少了室外庭院、阳台以及生活空间的日晒。

在三层的阳台上，通过倾斜的屋顶坡面，能够更好地欣赏周边的景色。为了应对新加坡当地的热带气候，在屋顶上又加了一层木质的镀层，适应多方位的角度，也可以减小从金属屋顶传入室内的热量。根据客户的要求，西方沿街立面的设计暗示了“房屋”的属性。但东立面经过精确的校准设计，为两间卧室以及共享的游廊和阳台提供极佳的视野。

通过利用在屋顶和立面挖出的有角度的天窗，太阳光可以散射或反射进入房间内部，照亮室内空间。同时，这些在建筑体量上或增或减的手法，不仅引入了自然光线，同时也营造出一种独属于这座房子的建筑语言，一种光和影的效果体验。

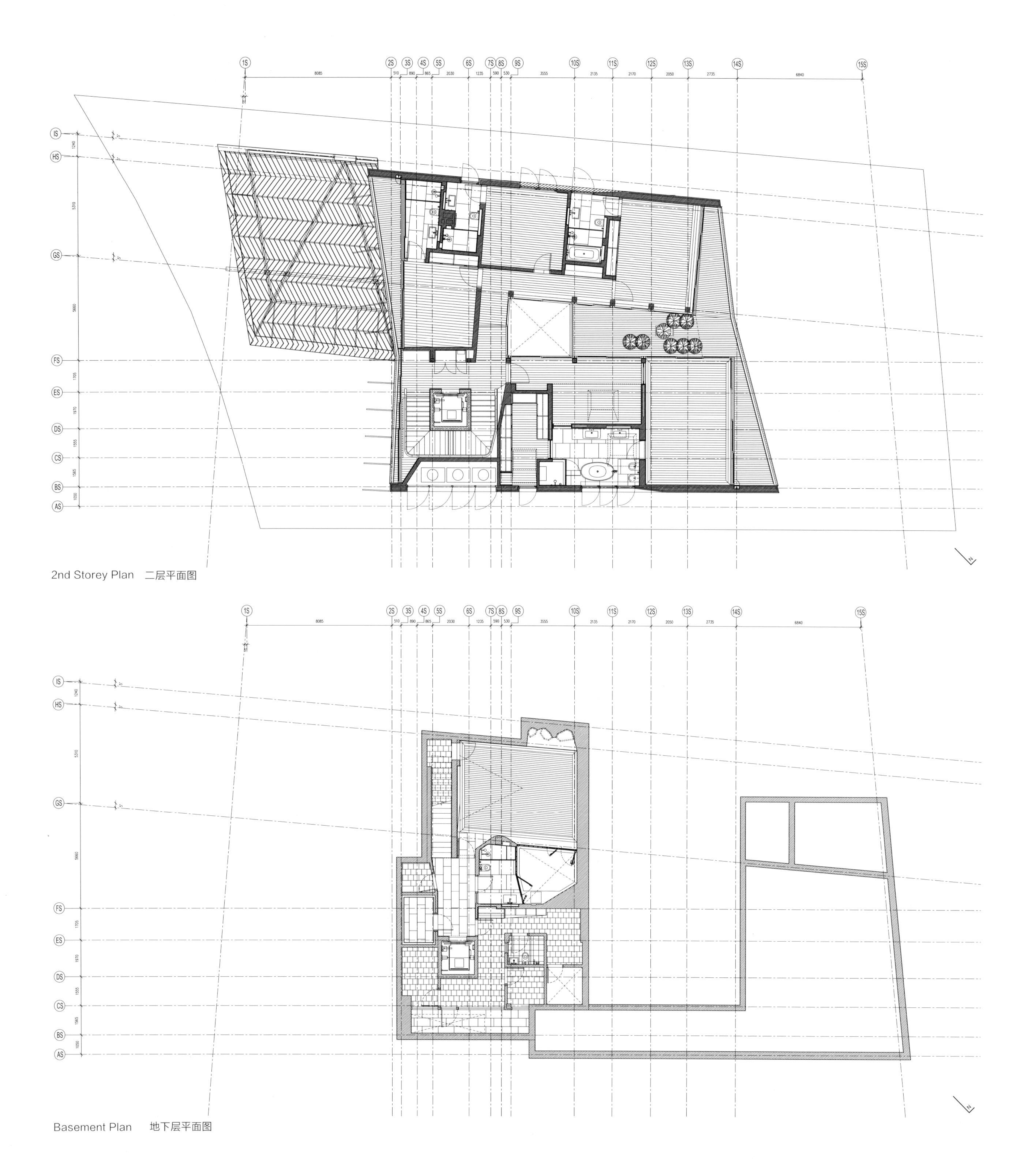

2nd Storey Plan 二层平面图

Basement Plan 地下层平面图

Section　剖面图

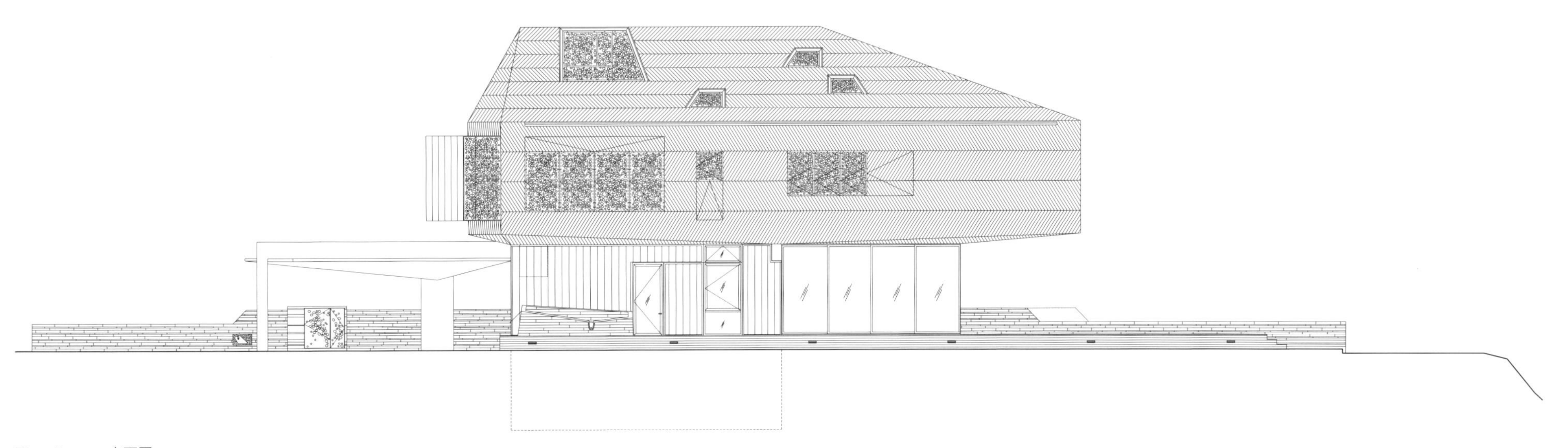

Elevation　立面图

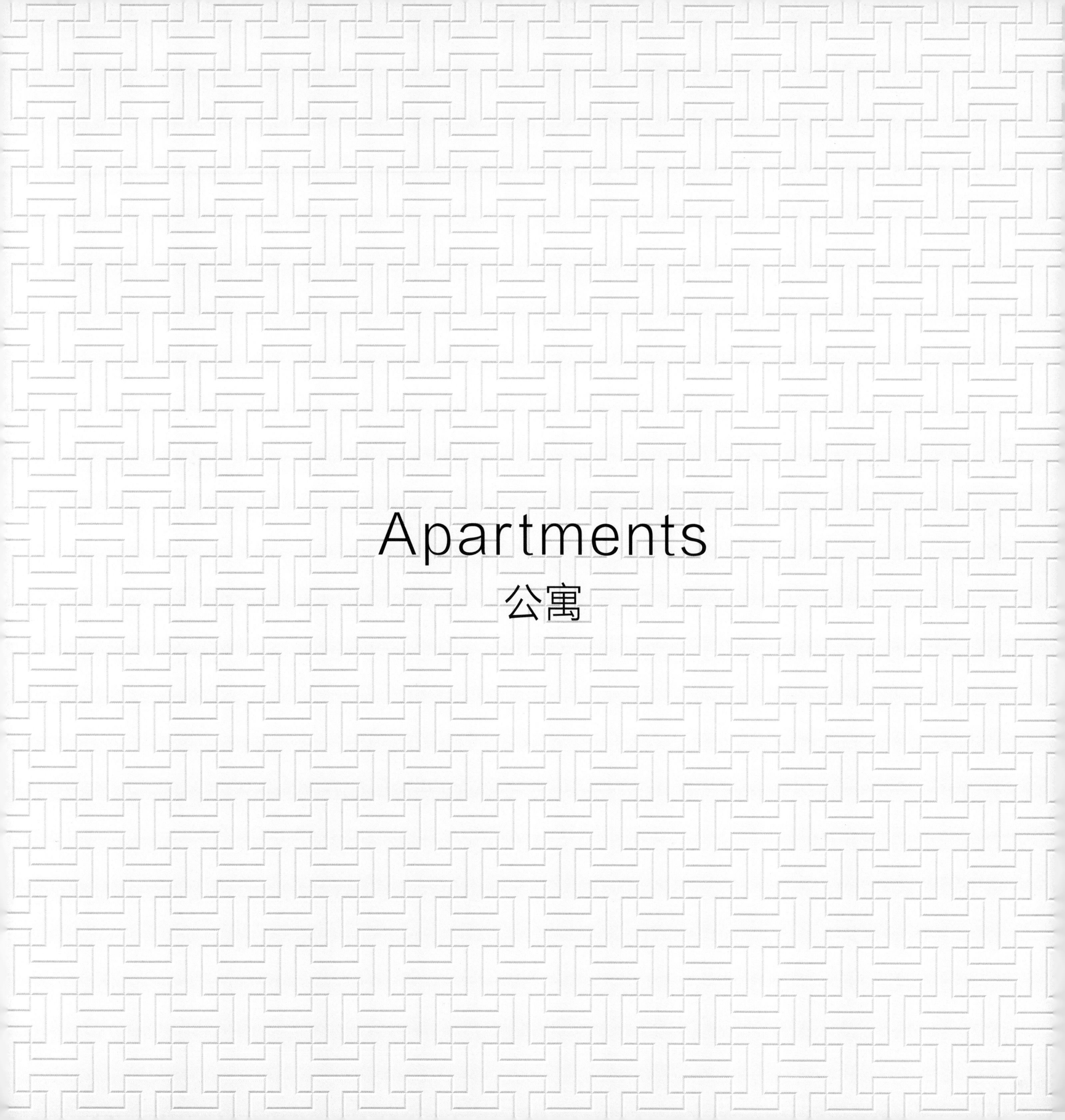

Apartments

公寓

Diversified Shapes　Open Space　Harmonious Atmosphere

形体多样　空间开放　氛围融洽

Location: Singapore

Client: Cove Development Pte Ltd

Architectural Design: DP Architects

Floor Area: 40,520m^2

项目地点：新加坡

开发商：Cove Development Pte Ltd

建筑设计：新加坡 DP 建筑设计事务所

建筑面积：40 520 m^2

Keywords 关键词

Vertical Extension 竖向延伸

Efficient Design 高效设计

Balance and Harmony 平衡和谐

Twin Peaks

双峰公寓

Overview

Standing proudly on the slopes of Leonie Hill, Twin Peaks towers are above the city, commanding views along Orchard Road and the Central Business District. It is a futuristic vision of a Singapore to come, where high-rise buildings embody the tenets of high-density, high-efficiency and high-end luxury lifestyle.

Twin Peaks is a product of Singapore's unique socio-political climate and is the first of a new generation of residences which truly caters to a contemporary city lifestyle. In designing this new typology, the architects of Twin Peaks have understood that today's urban dweller needs a home that is fully integrated with the city yet at the same time separate, and have provided a luxurious escape in which to unwind and enjoy the view.

Planning and Layout

The arrangement of the towers promises residents panoramic views of the Orchard shopping district, extending from the living room into the heart of Singapore. Twin Peak's position in the city centre means it is an ideal location for business and recreation with the main commercial areas only a stone's throw away. Residents will spend less time travelling and more time on the things they enjoy.

Architectural Design

Although situated on an expansive site, the architects decided to restrict the development's footprint and extend the towers vertically, leaving more space for lifestyle facilities. This resulted in the design of two symmetrical 35-storey towers that hold 231 units each. Not only is this a highly efficient design, but one of beauty that is balanced and composed within its urban context.

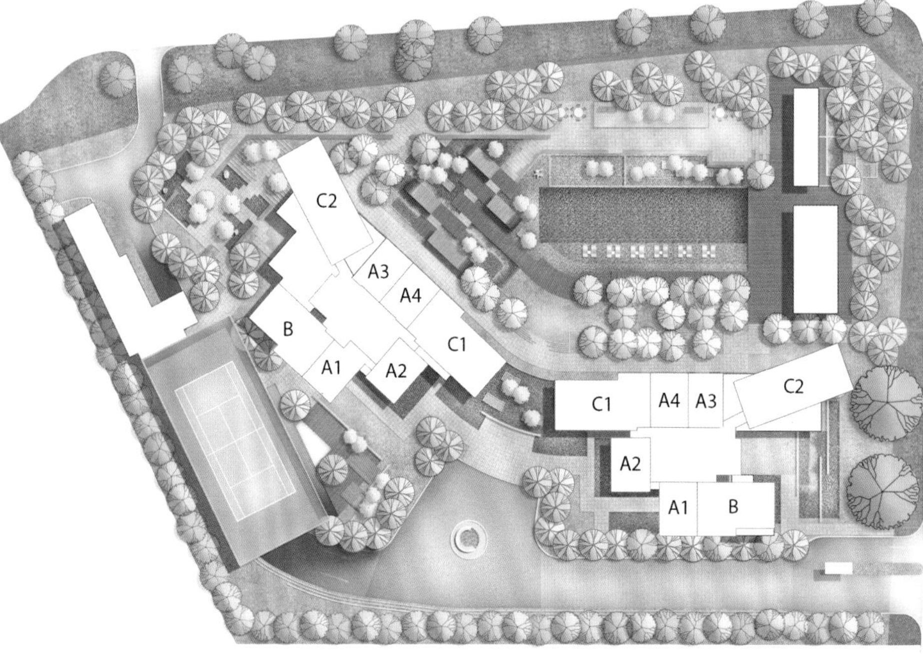

Supporting Facilities

To this end, Twin Peaks has many lifestyle facilities for residents to take pleasure in, including numerous pools, sky gyms, gardens and beautiful dining facilities. The importance of these amenities is expressed on the exterior where the vertical volume of the buildings is punctuated with the landscaped sky gym, creating a signature void.

There is also a dedicated concierge service on hand to support the residents' busy lifestyles. Residents may even enjoy the privilege of housekeeping and hospitality services extended by the Mandarin Orchard Singapore.

In Twin Peaks, every effort has been taken to make sure the residents are afforded the luxurious lifestyle they desire, one which begins as soon as they receive their new apartment. With this in mind, each unit comes fully furnished with high quality furniture and classic designer pieces.

项目概况

双峰公寓（Twin Peaks）坐落于新加坡的利安尼山路（Leonie Hill）上，傲立于城市之巅，俯瞰着从乌节路一直延伸到中央商务区的城市景观。双峰公寓为人们展现了新加坡的未来愿景：在此，高层建筑象征着高密度、高效率以及高端舒适的生活方式。

双峰公寓是新加坡独特的社会政治环境下诞生的建筑，作为新一代住宅的首个范例，真正地满足了现代都市人的生活方式的需求。在设计这一新型住宅的同时，双峰公寓的建筑师理解到当今的城市居民需要一个与城市完全融合但同时又独立的住所，于是建筑师提供了这样一种既可以欣赏城市景观又可以放松减压的专享空间。

规划布局

双峰公寓的布局使住户在住所内就可以欣赏到乌节路购物区的全景，仿佛将客厅一直延伸到新加坡的中心地区。又因为地处市中心，双峰公寓与主要商业区仅一步之遥，是理想的工作娱乐场所。因此，住户可以节约通勤时间并将更多的时间花在喜欢的活动上。

建筑设计

虽然双峰公寓位于昂贵的地段，建筑师决定限制项目的占地面积而竖向延伸塔楼，从而为生活配套设施留出更多空间。由此形成了两栋对称的 35 层高的塔楼，分别有 231 户住户。双峰公寓不仅是一个高效的设计，而且值得一提的是，其平衡和谐的外观与周围城市环境浑然一体。

配套设施

为此，双峰公寓为住户提供了很多可以增加生活乐趣的配套设施，包括多个游泳池、空中健身房、花园和环境优美的餐饮服务。这些设施的重要性同时体现在建筑外观上：与景观相结合的空中健身房创造了一个标志性的虚空间，从而丰富了塔楼的垂直体量。双峰公寓还设置了专属管家式服务以支持住户忙碌的生活方式。住户甚至可以享受到邻近的乌节文华酒店（Mandarin Orchard Singapore）所提供的酒店式接待服务。

双峰公寓的设计还尽其所能地确保住户从入住的那一刻开始就能享受到所期待的最舒适的生活方式。为此，每个单位都配备了齐全的高品质家具和设计师的经典作品。

Location: 1 Grange Garden, Singapore

Design Consultant: Guida Moseley Brown Architects

Design Partner: Harold Guida

Administrative Partner: Steve Moseley

Team: Michael Komnacki, Sieglinde Whittle, Andrew Fox

Architectural Design: Mr Choy Meng Yew ,Ms Phyllis Yiu (P&T Consultants Pte Ltd)

Landscape Design: Sitetectonix Private Limited

项目地点：新加坡 Grange Garden 路 1 号

设计顾问：Guida Moseley Brown 事务所

设计主创：Harold Guida

执行搭档：Steve Moseley

项目团队：Michael Komnacki, Sieglinde Whittle, Andrew Fox

项目设计：Choy Meng Yew, Phyllis Yiu(P&T 咨询有限公司）

景观设计：Sitetectonix 私人有限公司

Keywords 关键词

High Transparency 高度透明

Different Styles 风格迥异

Upward Views 仰视景观

Colourful Stairs 彩色阶梯

The Grange

The Grange 公寓

Design Concept

The design of the Grange originated in understanding the special location within the city and the potential for outward views in every direction. Like spreading tree branches seeking the sunlight, the plans of the two towers have been organized to provide each unit with maximum light and views. The unique plans, of three units per floor, have been conceived so as to place the living rooms of every unit at the outward, all-glass perimeter. These highly transparent spaces have views on three sides. To enhance this plan, the towers have been lifted onto columns to enjoy higher and longer views, above the tree canopy of adjoining estates, but looking down onto this green foreground. The combination of height and transparency results in a special architecture.

The spreading plan configurations of the two towers establish a dynamic presence that creates an interesting relationship between the buildings and sets out a genesis of diagonal view lines for the development of the landscape.

Landscape Design

The open character of the towers at the lower five levels allowed for the landscape design to originate under the towers and down at the car park level, employing plantings and reflecting pools to create a welcoming light-filled environment for each individual lift lobby. At pool – plaza level, bridges provide access for visitors to the upper levels of these glass-sided lobbies. From here the landscape expands into the whole of the site, developing into differing types of gardens and individual features designed especially for this place, and providing for choice of environment and activity for the residents.

Recreation Design

As well as the landscape, the recreation club is developed in the columned spaces and provides a range of meeting and health facilities, including exterior roof terraces overlooking the pools and gardens. The club and the various lobbies are interconnected to a vehicular drop-off provided with a large glass-roofed cantilevered canopy by way of similar covered walkways, and special coloured lighting is integrated into these structures enhancing the direction of movement at pool—plaza level.

Towers Design

The towers, composed of three and four bedroom units, are designed to have a striking form and character, with strong horizontal bandings that occur on alternate floors establishing a secondary, two-storey scale within the towers that are slim, unified, and unbroken when viewed toward the glassy living rooms. Projecting balconies extend the outlook and provide an interesting serrated profile when seen against the sky. Many of the lower level architectural and structural elements were designed to contribute to composing interesting upward views, an aspect important in defined environments such as this.

The two towers are set at different spacing from the pool – plaza level and reach different ultimate heights adding visual and experiential interest. At the top of the towers, penthouse units are marked with two-storey terraces with master bedroom balconies, and projecting stairs that create a unique experience for the occupants and form a strong sculptural and colourful terminus for the towers. These highly visible elements have been designed to be seen as special markers in the Singapore skyline, and the stair colours are "drawn-down" within the columned space adding vibrancy and a sense of unity to the towers. This aspect is further developed with lighting that engages the columned space with the glazed lobbies, the landscape, and the colours of this highly visible aspect of the development.

Ground Level 一层平面图

Second Level 二层平面图

Second Level 二层平面图

Floor Plan Layout 屋顶层平面图

设计理念

Grange 项目的设计理念起源于它独特的地理位置以及潜在的多方位室外景观。方案设计中，两栋楼都最大限度地留给彼此足够的光照和视野，正犹如两支努力向天空伸展的树枝。该项目的独特构想是每层设置三个单元，每个单元的客厅都朝外以创造出一个全玻璃的建筑外围，这样高度透明的空间也使住户能全方位欣赏到三个方向的景观。为此，设计师们将住宅楼的底柱抬高至高于楼盘周边的树冠，使得住户能得到更高更远的景观视野，也能俯瞰楼盘前坪的绿地。这种高度和透明度的组合，彰显出了楼盘的特别之处。

两栋塔楼的配置营造出充满活力的氛围，使得它们之间出现了有趣的联系，并且为小区的景观设计提供了一条对角视线。

景观设计

该楼盘的一至五层所具有的开放性使得景观设计能延伸至停车场，利用植被和池塘的反射为每个电梯门廊营造出一个热情的且日照充足的环境。在与广场齐平的池塘周边设立的一些小桥，能让游客轻松往来于拥有玻璃墙面的大堂。从此处延伸开来，逐渐发展成风格迥异和具有独特功能的景观，为住户提供了众多活动和休闲的选择。

娱乐会所设计

同景观设计一样，娱乐会所设置在一个圆柱形空间里，拥有露天的屋顶平台可俯瞰庭院的池塘和花园，可满足各种会议和健身的需要。通过各个相似的走廊，该会所和众多大堂与有着悬挑玻璃顶的汽车下客点相连，再加上各色灯光的融合，能指引宾客前往池塘广场方向。

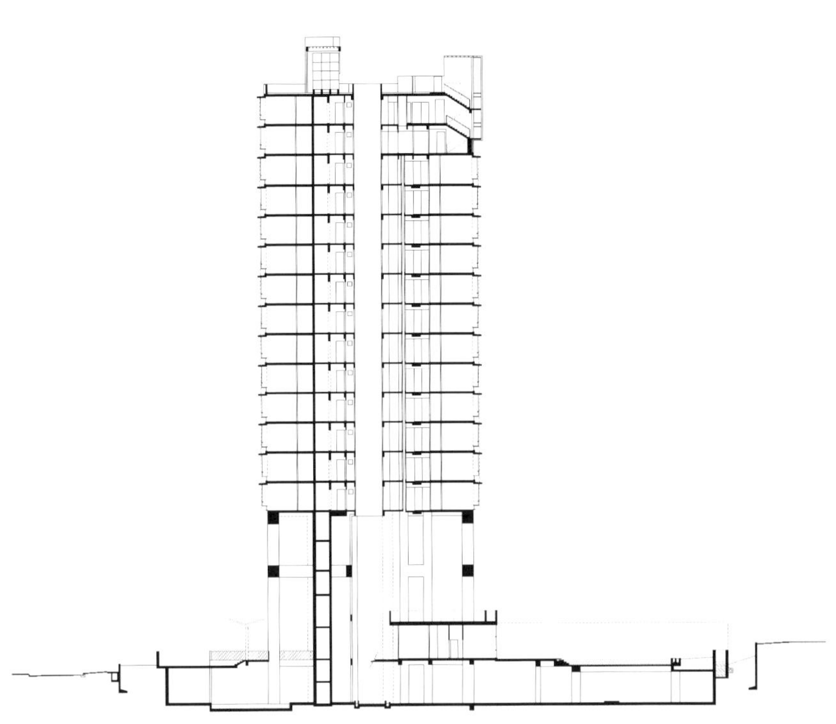

Section　剖面图

塔楼设计

分别拥有 3 房或 4 房户型的两栋塔楼，在建筑物外设计有坚固的交替水平条状带，当视线扫至楼盘玻璃面的客厅时，映入眼帘的是一个两层楼高的标度，纤细完整而且统一，这一切都旨在使其拥有一个醒目的楼盘外形和特性。向外凸出的阳台不仅为住户扩展了视野，当望向天空的时候，它们便又构成了一道有趣的齿轮状外形轮廓。与此相同的是，低楼层中许多的建筑要素和结构要素，在设计阶段便是为了能形成新颖有趣的仰视景观，以得到一个更好的外形轮廓。

两栋塔楼拥有不同的池塘广场间距和楼高，这增加了住户的视觉和体验的兴趣感。在楼盘的顶层公寓，在一个两层楼高的露台上设计有主卧室和凸出的楼梯，为住户提供了独特的体验，同时也为塔楼创造了一个多彩的雕像般的顶端。这些高度可见的元素被视为新加坡天际线上一道特殊的痕迹。并且，那些从圆柱空间里垂直下来的彩色阶梯也为塔楼增添了一丝活力和整体感。在灯光的照射下，圆柱形空间内的大堂以及景观，都愈发的闪耀夺目了。

Location: Singapore

Contractor: Thian Sung Construction Pte Ltd

Architectural Design: Guida Moseley Brown Architects

Landscape Design: Sitetectonix Private Limited

Design Team: Harold Guida, Michael Komnacki , Michael Liu

Structural Engineer: DE Consultants Pte Ltd

Electrical Engineer: Rankine & Hill (S) Pte Ltd

Mechanical Engineer: Rankine & Hill (S) Pte Ltd

项目地点：新加坡

承包商：宋天建设有限公司

建筑设计：澳大利亚 Guida Moseley Brown 建筑设计事务所

景观设计：Sitetectonix Private Limited

设计团队：Harold Guida, Michael Komnacki , Michael Liu

结构工程：DE 咨询有限公司

电器工程：Rankine & Hill (S) 有限公司

机械工程：Rankine & Hill (S) 有限公司

Keywords 关键词

Y Shape "Y"字形

Glass Curtain Wall 玻璃幕墙

Sky Terrace 观景台

Helios Residences

嘉旭阁

Overview

Helios Residences is located in one of Singapore's most densely developed residential areas, Orchard Road, a precinct including towers and compactly positioned terrace houses, all within close walking distance of the linear retail and entertainment heart of the city. Three interlinked towers "zig-zag" into the "Y" shape and the first occupied floor is one level above the upper road level, leaving the site to step down with terraces, gardens, and courts, with car parking structures below these various amenities. This concept is further enhanced by maximizing the building height to leave as much open space at the garden levels as possible.

Planning

Resident amenities are positioned at storey 4, a double height "sky terrace" and an independently supported swimming pool deck extends from this level, 15 metres above a garden at the level of vehicular drop-off. These common amenities include a gym with windows overlooking the entrance sculpture court below, a children play garden, change rooms with steam and sauna rooms, and a variety of landscaped places providing differing character to support different types of social activities.

Architectural Design

The structural frame is reinforced concrete with brick infill walls and partitions. The external wall finish is long-life paint on render, cement plaster internally. The building is substantially enclosed in custom designed curtain wall incorporating high performance low-e glass, and glass blocks at the spandrel panels with integrated LED lighting. The curtain wall system also incorporates aluminium louvers at the air conditioning condenser unit ledges.

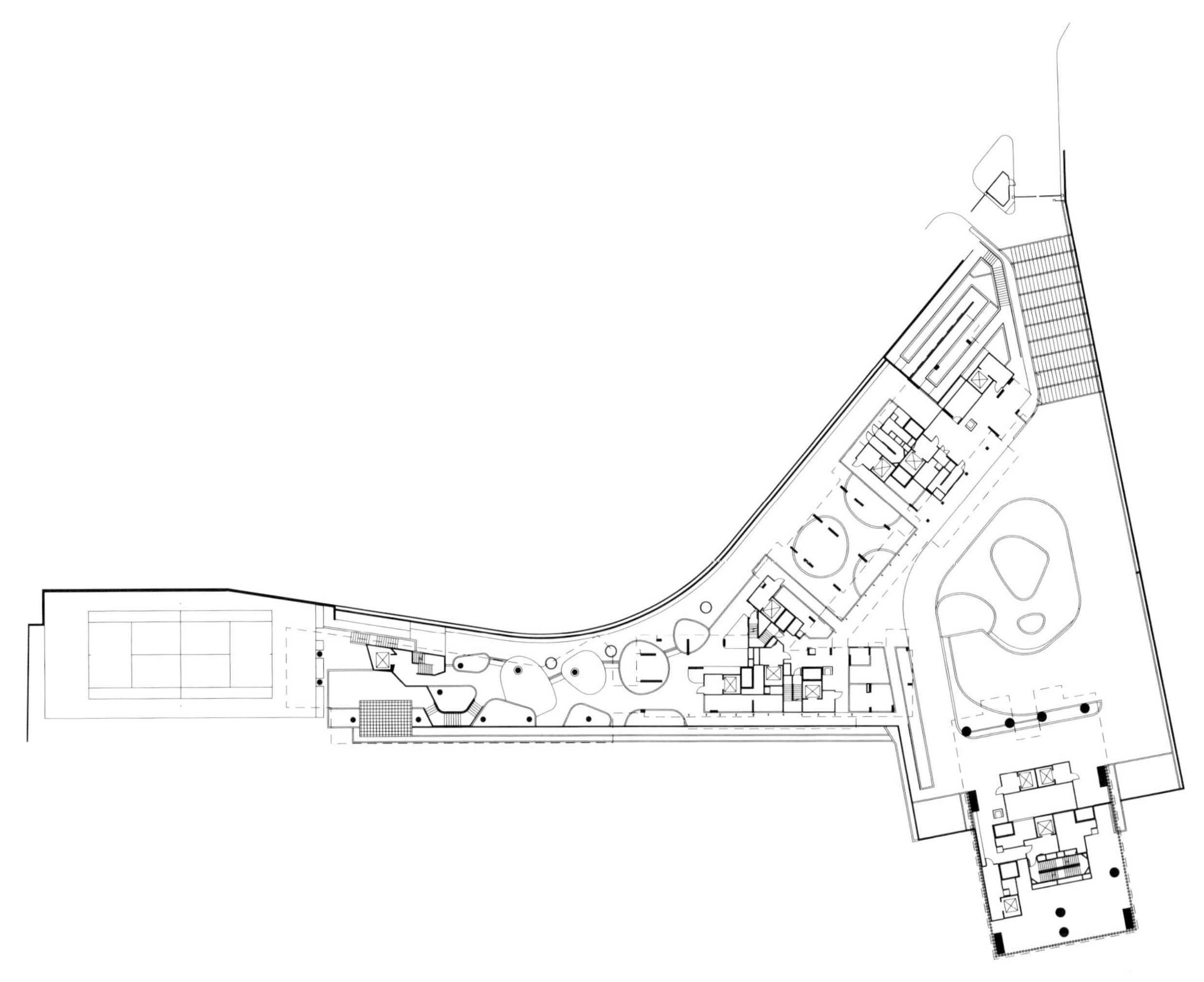

Basement　　地下层平面图

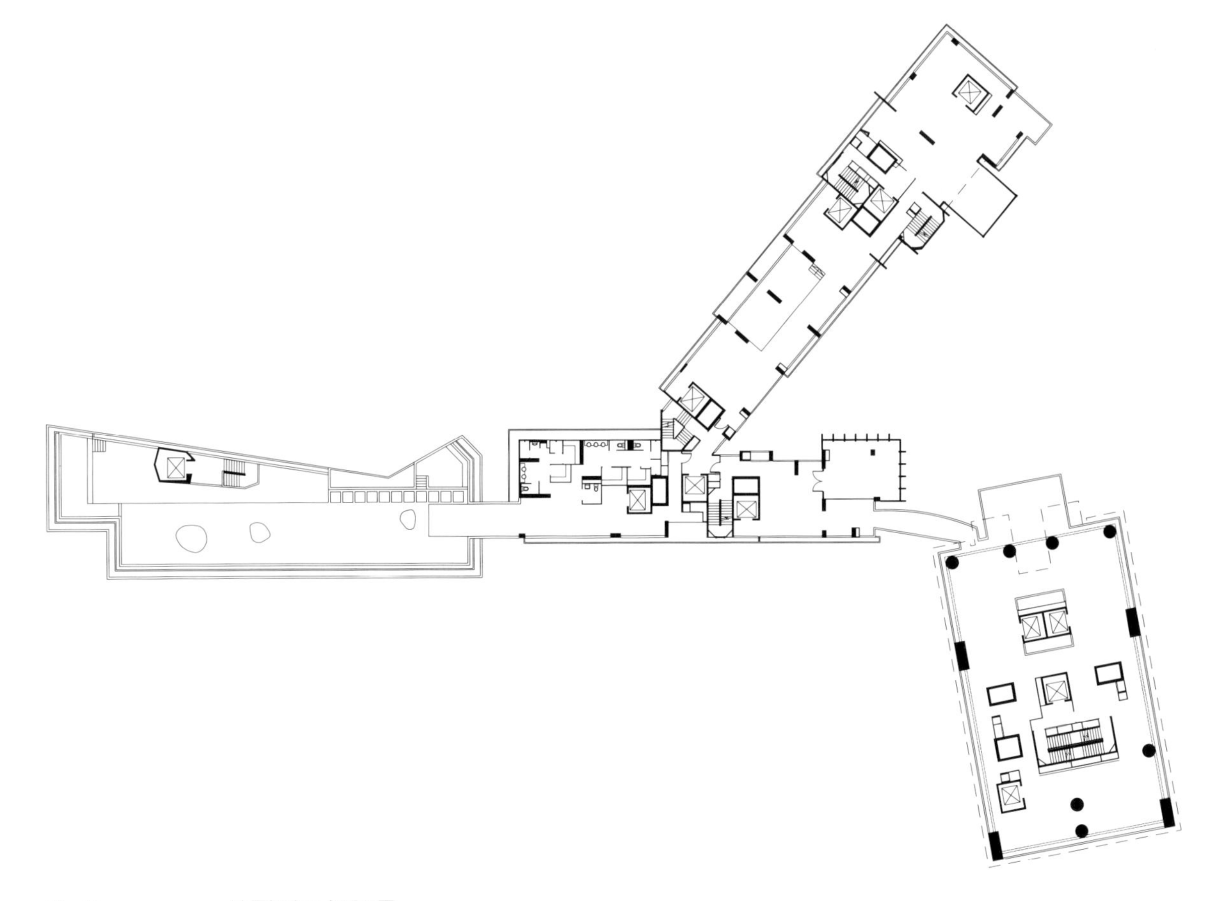

Sky Terrace　　凌霄阁摩天台平面图

项目概况

嘉旭阁位于新加坡住宅区最稠密的地段——乌节路，项目包括塔楼及洋房，步行即可达市中心的购物中心及娱乐中心。三栋相互连接的塔楼被设计成"Y"字形，首层为其共有楼层，比路面高出一级，从而为露台、花园及广场预留了空间，停车场位于地下一层。设计理念通过提升建筑高度得以加强，并为花园预留出尽可能多的空间。

建筑规划

大楼从四层起即为住宅，设计师于此设计了一个双高的观景台及独立的延伸出去的泳池甲板，此处距花园的垂直距离为 15 m。此外还设计了可俯瞰雕塑广场的健身房，儿童游乐场地，带更衣室的桑拿房，以及一系列景观小品以满足不同性格住户的活动方式。

建筑设计

建筑结构构架为钢筋混凝土，墙内及分割处填充砖头。外墙喷涂一层高品质油漆，内墙为水泥泥灰。建筑外覆盖着定制的高性能低辐射玻璃幕墙，拱肩镶板处的玻璃块为集成 LED 照明系统。幕墙系统也包括空调冷凝部分的遮光栅格。

Level 6, 8, 10, 12, 14, 16 and 18　　6、8、10、12、14、16、18 层平面图

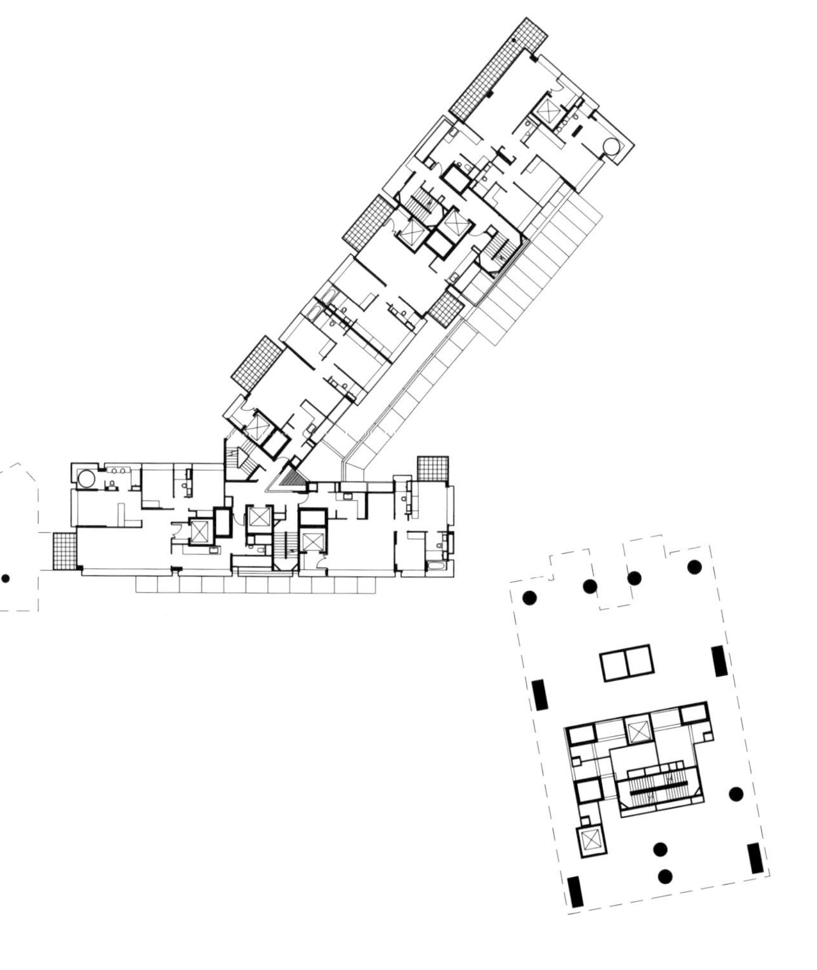

Level 1, 2 and 3　1、2、3 层平面图

HELIOS

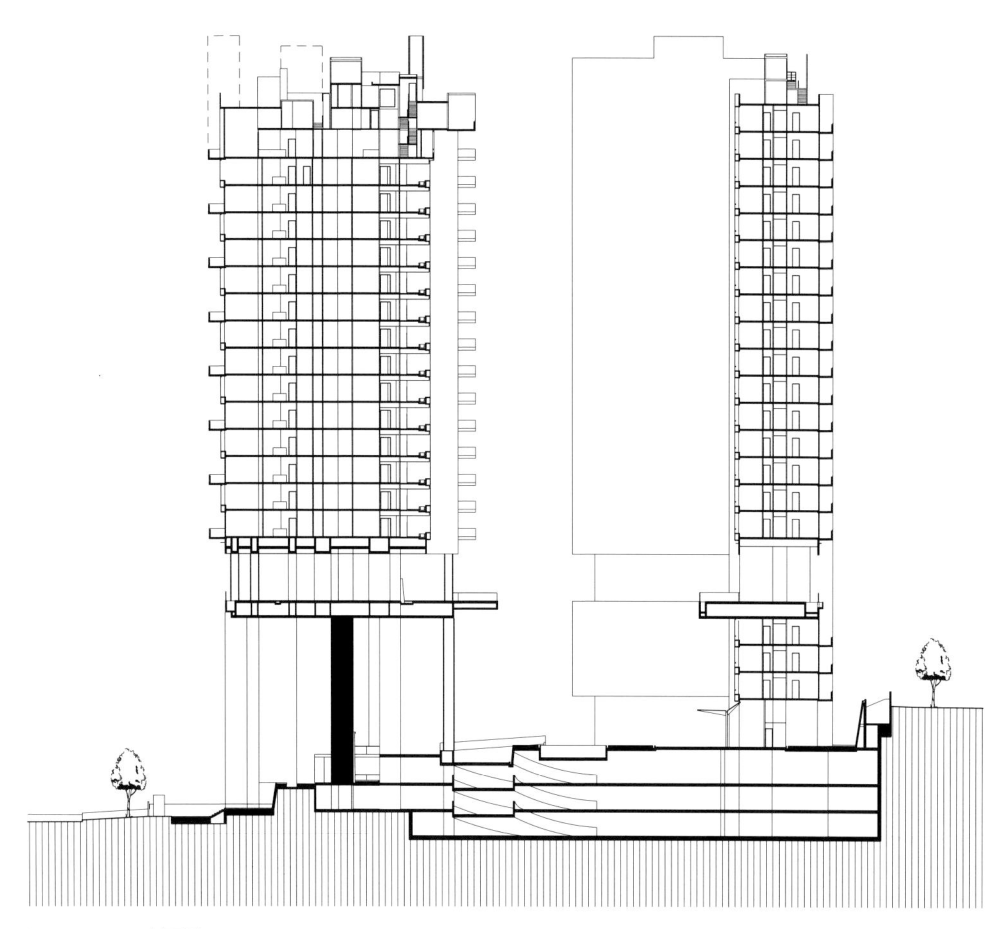

Section B 剖面图 B

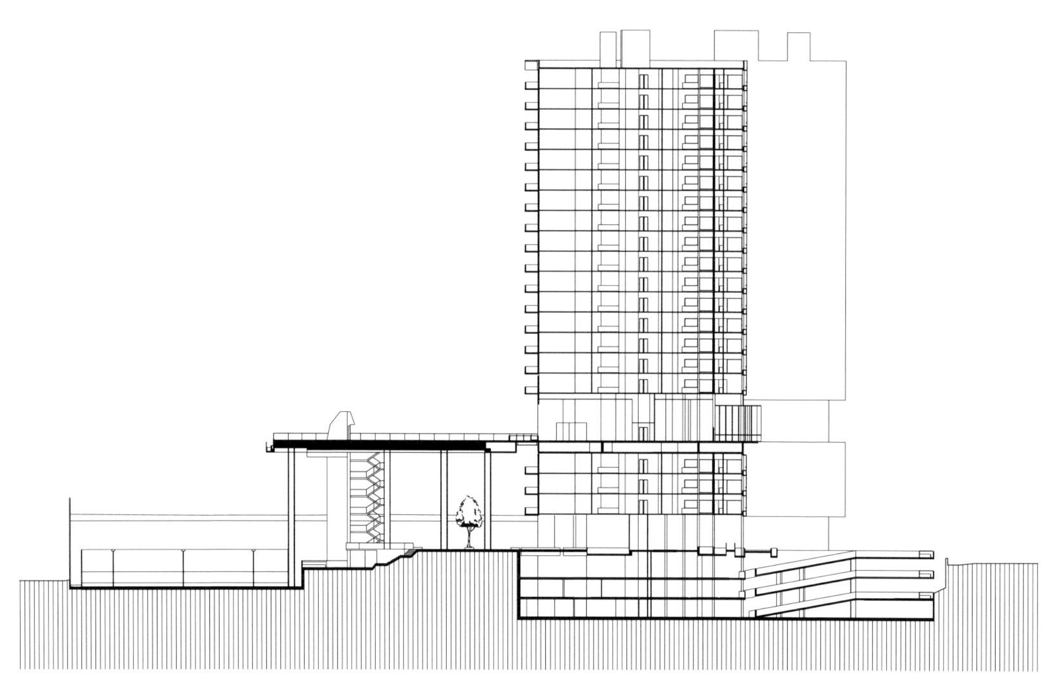

Section D 剖面图 D

0.9m

Location: Bukit Timah Road, Singapore

Architectural Design: WOHA Architects

Architect In Charge: Richard Hassell, Wong Mun Summ

Floor Area: 69,350 m²

Photography: Patrick Bingham–Hall

项目地点：新加坡武吉知马路

建筑设计：WOHA 建筑事务所

设计主管：Richard Hassell, Wong Mun Summ

建筑面积：69 350 m²

摄影：Patrick Bingham-Hall

Keywords 关键词

Spacious Courtyard 宽敞庭院

Living Experiences 居住体验

Green Living 绿色环保

Goodwood Residence

优景苑

Overview

Set off the prime Bukit Timah Road area, and against a verdant 20-hectare(200,000 m²) backdrop of greenery, the 210-unit Goodwood Residence is conceived on a macro scale as a breathing space—a rarity in high density urbanized Singapore—and an extension of the Goodwood Hill tree conservation area that it shares a boundary of 150m with. Articulated as two 12-storey L-shaped blocks, the 2.5-hectare(about 25 000m²) development dialogues with the hill that it embraces and merges with in a language of openness and continuity made expressive by varying degrees of scale and privacy.

Courtyard Design

All units are one apartment thick, with the blocks configured like boundary walls that define and enclose a series of courtyards that first draws its residents through an intimately scaled tree-lined boulevard that wraps around a pair of foliage screened tennis courts, leading to a formally scaled cobblestone entrance courtyard, into an expansive central open lawn/swimming pool. This spacious courtyard that visually merges into Goodwood Hill, measures some 100m across opposite blocks, which enhances the privacy of its residents while offering excellent views towards the lush greenery. It is the main community gathering place and breathing room of the development, complete with club house facilities (concierge, reading lounge, private function room and pools) that parallel that of serviced apartments. Smaller landscaped courtyards branching off the sheltered walkways are further extended into the basement carpark as entry points, making pleasant the homecoming experience with natural day light, fresh air and planting.

Floor Design

On a building scale, distinct strata of living experiences are crafted. The ground floor units are designed as a new typology of "landed housing apartments", with lofty ceilings, generous outdoor pool terraces and specially devised auto-sliding gates. Overlooking the central courtyard on the 2nd and 3rd storeys, are designed with treehouse cabanas perched amidst the treetop canopies, immersing its inhabitants in close-range nature. Rising above this tree line are the mid-levels (4th-11th storeys), which have overlapping double volume balconies. This culminates in the 12th storey penthouses that are sky-bungalows complete with generous roof/pool terraces that effectively recreate a new ground level with the added advantage of unobstructed city views and cooling breezes at elevated height.

Goodwood
RESIDENCE

Green Design

Inspired by patterns of traditional Asian woven textiles, all typical apartment units (2nd storey upwards) feature fine aluminium fins orientated at 45 degrees to north-south, that are devised as operable facade screens which not only provide vertical sun shading without compromising on ventilation, but also allows user-controlled amounts of privacy as well as facade animation. Planters of 1m width, coupled with projecting balconies of either 2.7m or 4.5m depth, further provide vertical greenery and horizontal shading for the apartments below.

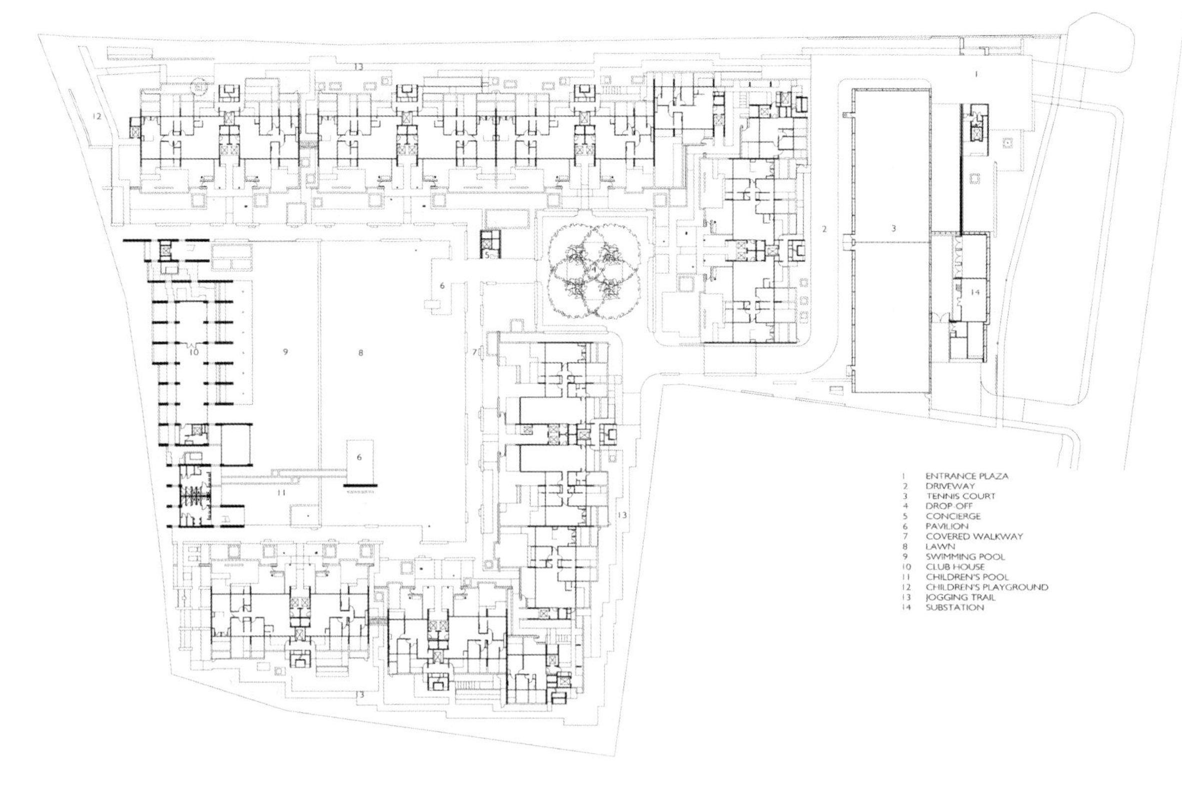

Site Plan 总平面图

项目概况

优景苑位于武吉知马路段，背靠良木山保育区约 20 公顷（200 000 m²）的茂盛雨林，在新加坡这样一个高度都市化的地区，拥有 210 个单位的优景苑无疑为住户提供了一个很大的休息空间。项目占地面积大约为 2.5 公顷（约 25 000 m²），共有两栋 12 层高的 L 形住宅楼，通过规模大小和隐私空间的变幻，该住宅小区的开放、连绵不断的建筑与良木山交相呼应。

庭院设计

小区住宅均为一梯两户。住宅楼的边界墙围出了许多庭院，住户穿过一条环绕着网球场的林荫车道，便来到了入口处的庭院。进入小区，则是一片广阔的草地和泳池。这个宽敞的庭院最宽处可达 100 m，并与良木山交相呼应，在保护住户隐私的同时，也为他们提供了迷人的绿色景观。该庭院是小区内主要的社区休闲场所，包括各种俱乐部会所设施（如礼宾室、阅览室、私人活动室以及泳池）。从小区内的行人通道分叉出了一个较小的景观庭院，能直达地下车库，使住户开车返家时能享受到新鲜的空气和怡人的环境。

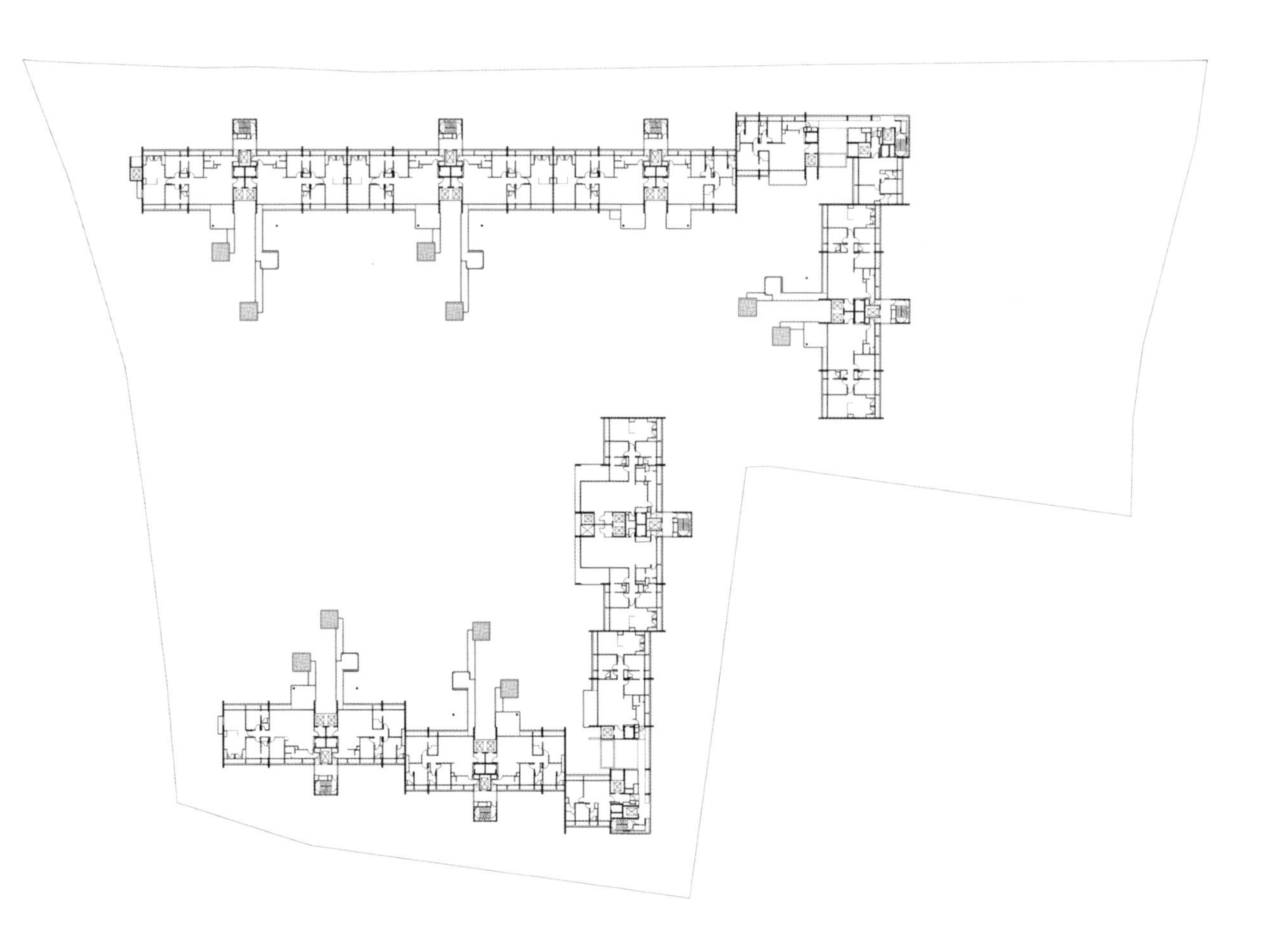

3rd storey plan 三层平面图

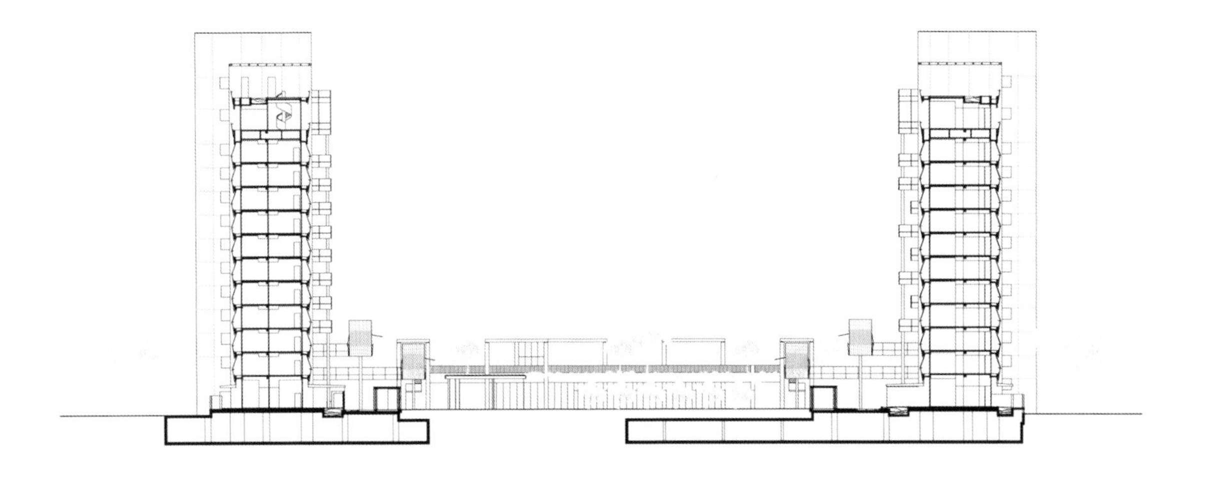

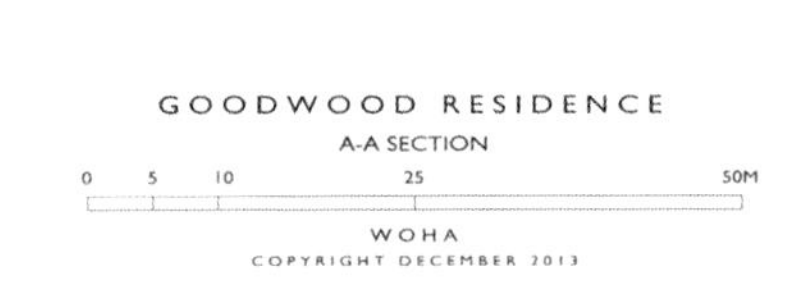

Section A-A　　A-A 剖面图

Section　剖面图

Section　剖面图

楼层设计

该小区为不同楼层的用户创造了不同的居住体验。位于一楼的单位重新阐释了底层公寓：挑高的天花板，宽敞的户外泳池露台，以及专门设计的自动滑动门。而二楼和三楼不仅能俯瞰中央庭院，而且在树顶上建筑了一个小屋，使住户能更加亲近自然。再往上便是拥有错落有致的双倍大阳台的中层住宅层（4 楼至 11 楼）。在 12 楼的顶层住宅设计成平房，为住户提供了硕大的屋顶和泳池露台，使他们能毫无遮挡地享受到城市风光和清凉徐风。

绿色设计

受传统亚洲纺织品图案的启发，该小区所有的传统户型（二楼及以上）在南北偏移 45° 方向安置了铝制隔热片。这些隔热片不仅在保证通风的前提下能防止日晒，而且能随意调节以保护隐私，也能使建筑立面变幻不一。在深约 2.7 m 或 4.5 m 的突出阳台上设置了 1 m 宽的植物带，在延伸了绿色景致的同时，也为阳台下方的住宅提供了更多阴凉。

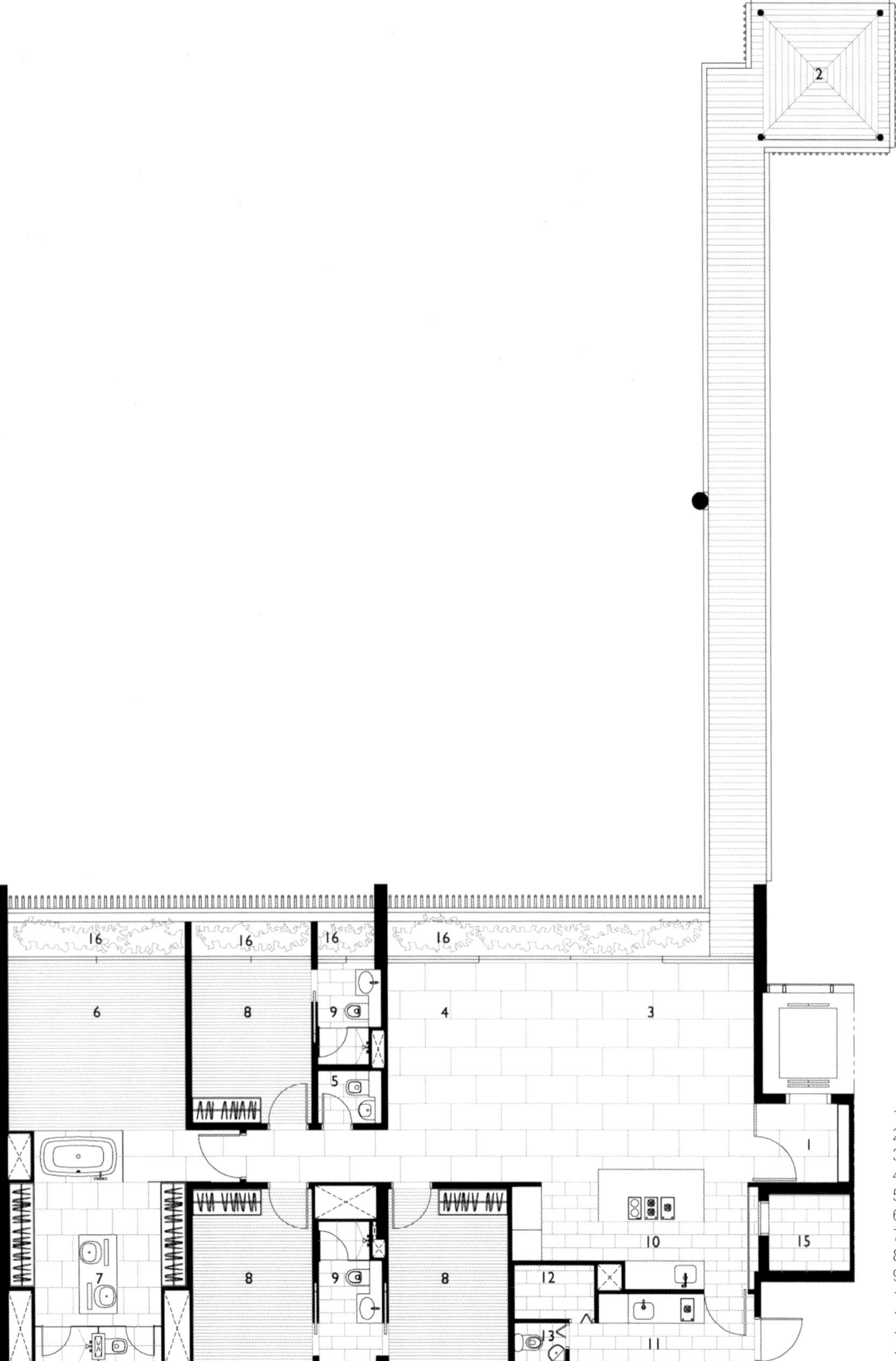

Bedroom Unit Plan　卧室的单元平面图

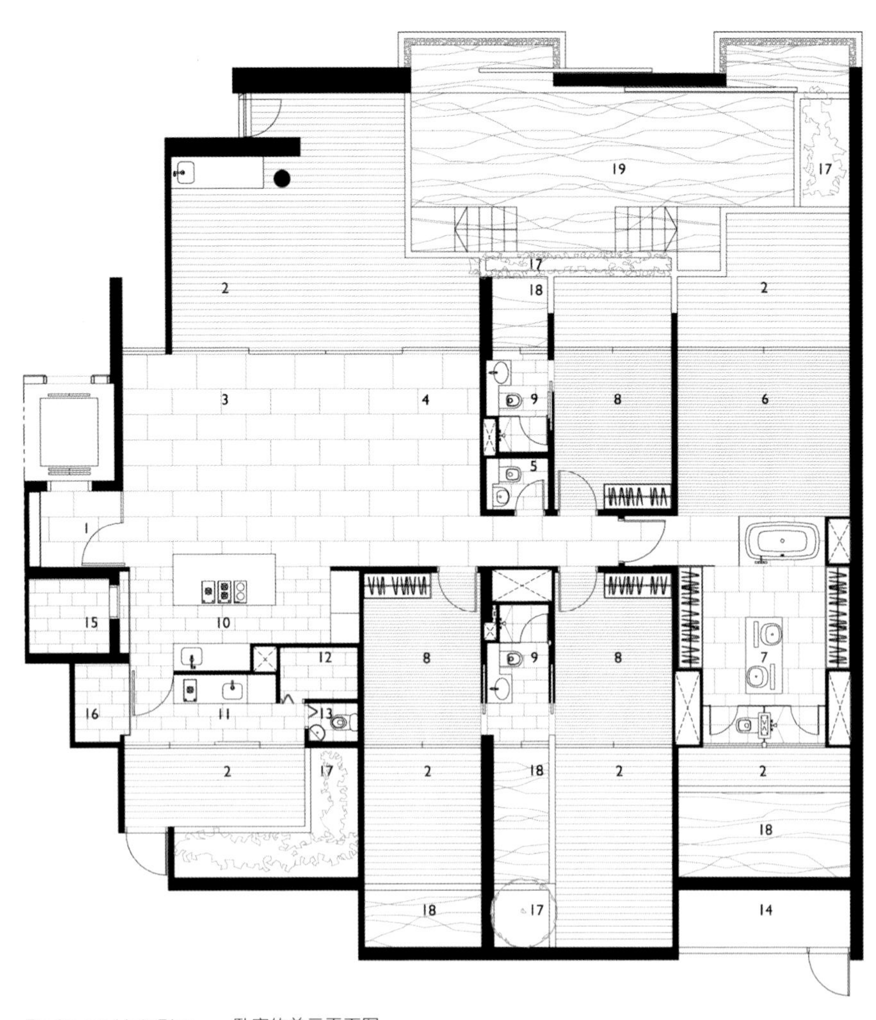

1.Private Lift Lobby	1. 私人电梯厅
2.Pes Deck	2.PES 平台
3.Living	3. 客厅
4.Dining	4. 餐厅
5.Powder Room	5. 化妆间
6.Master Bedroom	6. 主卧
7.Master Bathroom	7. 主卧浴室
8.Bedroom	8. 卧室
9.Bathroom	9. 浴室
10.Dry Kitchen	10. 厨房干区
11.Wet Kitchen	11. 厨房湿区
12.Utility Room	12. 功能用房
13.Bath	13. 浴室
14.Ac Ledge	14. 空调架
15.Household Shelter	15. 遮蔽区
16.Storage	16. 储物间
17.Planter	17. 植物
18.Reflection Pool	18. 镜面池塘
19.Swimming Pool	19. 泳池

Bedroom Unit Plan　卧室的单元平面图

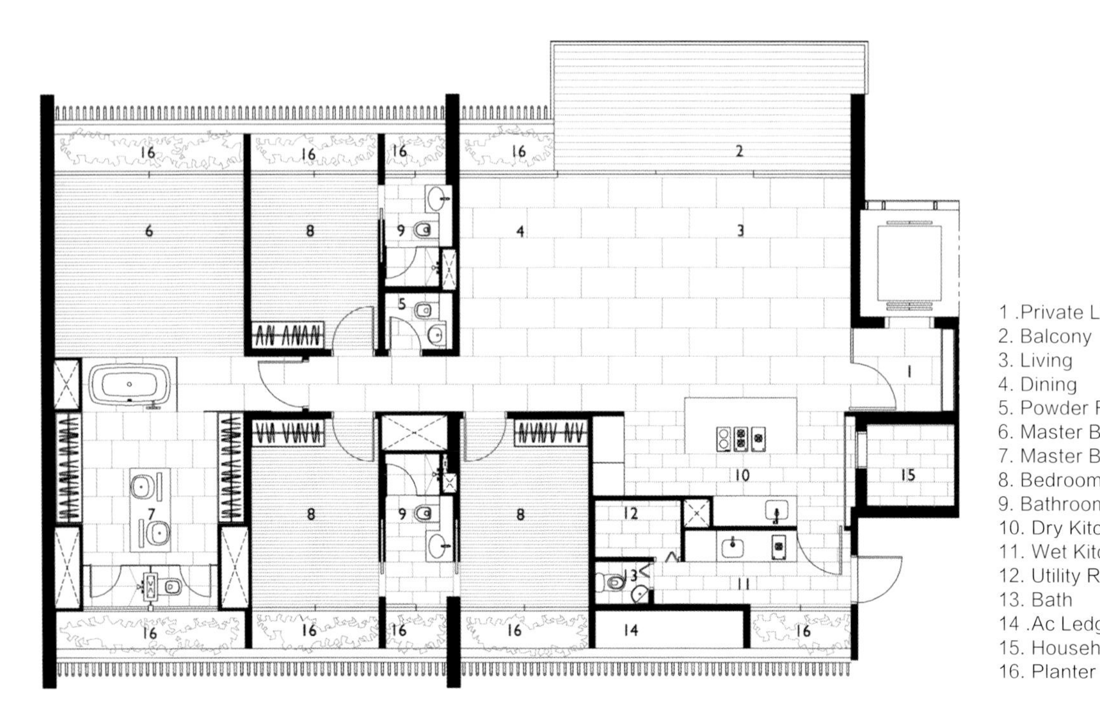

1 .Private Lift Lobby	1. 私人电梯间
2. Balcony	2. 阳台
3. Living	3. 客厅
4. Dining	4. 餐厅
5. Powder Room	5. 化妆间
6. Master Bedroom	6. 主卧
7. Master Bathroom	7. 主卧浴室
8. Bedroom	8. 卧室
9. Bathroom	9. 浴室
10. Dry Kitchen	10. 厨房干区
11. Wet Kitchen	11. 厨房湿区
12. Utility Room	12. 功能用房
13. Bath	13. 浴室
14 .Ac Ledge	14. 空调机架
15. Household Shelter	15. 遮蔽区
16. Planter	16. 植物

Bedroom Unit Plan　卧室的单元平面图

Location: Bukit Timah Road, Singapore

Developer: Orwin Development Limited

Architectural Design: DP Architects

Land Area: 44,400 m^2

项目地点：新加坡武吉知马路

开发商：Orwin 发展有限公司

建筑设计：DP 建筑师事务所

占地面积：44 400 m^2

Keywords 关键词

Corner Timber Treatment 木质材料

Clean Solid Lines 明朗线条

Garden Terrace 花园露台

Floridian

Floridian 豪华公寓

Overview

As a significantly sized residential development along the prime Bukit Timah Road, the Floridian defines a new type of residential living, one of coming home to the hotel and celebrating the arrival as part of the everyday coming home. The ground level is planned to be free of vehicular traffic as much as possible and instead filled with the rich interplay of lush greenery, recreational landscape and water features. The extensive water bodies along the main spine of the development invite waterfront living, with ground floor units enjoying the luxury of water lapping right outside their living and dining spaces.

Architectural Design

The Keys (Towers 3 & 4) characterizes the entrance with the corner timber treatment at the communal green platforms on alternate levels upon arrival. Adjacent to the lap pool, the Downtown towers (Towers 1 & 2) at the front of the development plays up an urban quality of clean solid lines and accent colours as a reflection of the young urbanites' lifestyle.

Aligned next to the Keys is the Verandah (Towers 5, 6 and 7) which is distinctive because of the communal rooftop terraces and gardens that embellish the facade and roofscape with lush green foliage. The Verandah is unique with balconies at the front and garden terrace at the kitchen, providing residents with alternatives for living and dining at either frontage.

The highest tower of the Verandah cluster is Tower 8, which stands at the highest level of the site and commands a grand vista of the central waterscape. Similar to the other Verandah towers, these units have double frontage and enjoy breezy cross ventilation.

The Hideaway and Pavilion Towers (Towers 9, 10 & 11) house most of the prime four and four-plus bedroom units and are defined by the generous, warm- schemed screened living/ dining balconies and corner balconies that one can walk out from the master bathroom. The Hideaway towers (Towers 9 and 10) are complemented with a SPA sanctuary at the ground level. The Pavilion tower (Tower 11) stands out gleaming in the vast main pool and enjoys unobstructed view of the waterscape and surrounding woods that stand amongst the good class bungalow of the area.

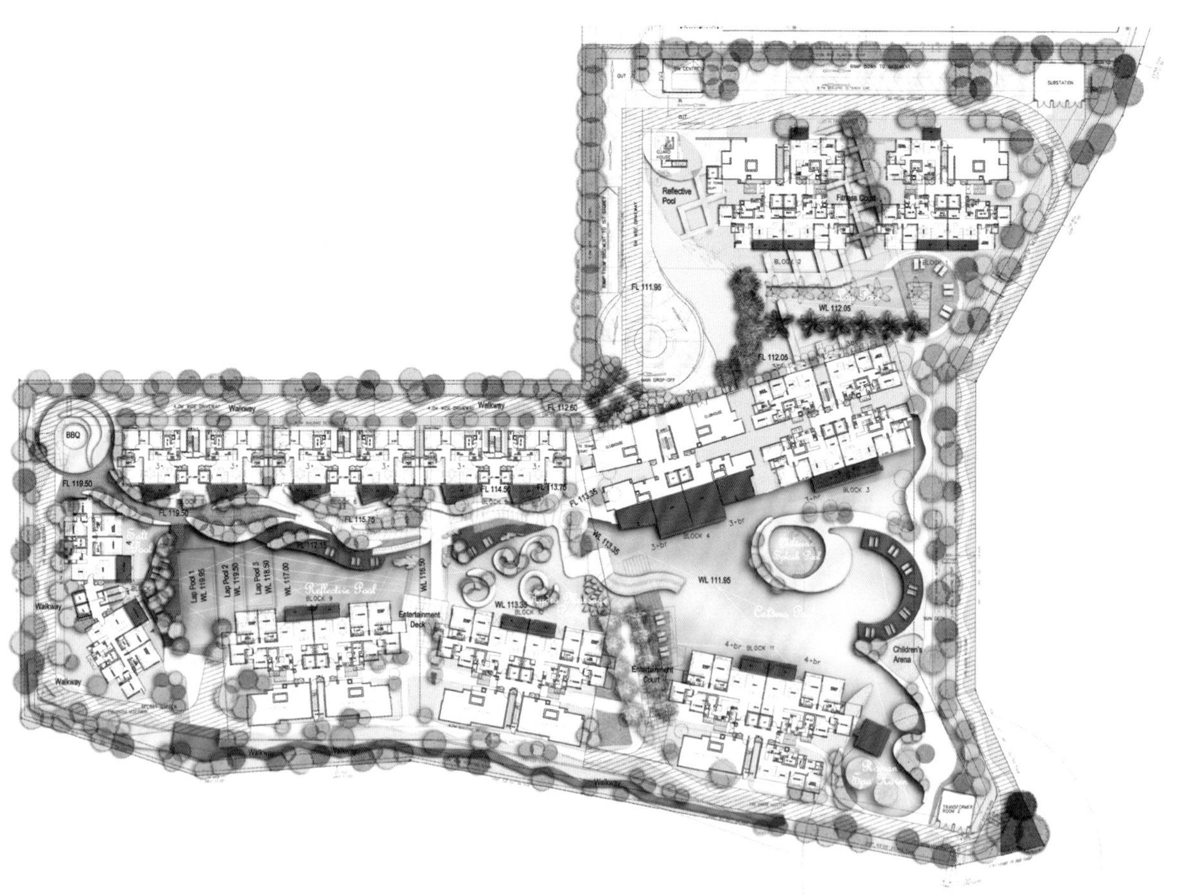

项目概况

作为武吉知马路上备受瞩目的住宅项目，Floridian 豪华公寓重新定义了居住体验，使住户回到家就像踏进了豪华酒店一般。地平面为了最大限度地杜绝交通污染，布置了大量的茂盛植物、休闲景观以及水景。贯穿整个项目的河流使住户临水而居，在一楼单位的客厅和餐厅的右方即可欣赏到迷人的水景。

建筑设计

该项目的 Keys 楼（第 3、4 栋）在面向公共绿化平台的这边的入口处，每两层会有木质材料装饰在墙角。紧邻游泳池的 Downtown 楼（第 1、2 栋）布置在最前面，用明朗的线条和色彩来反映年轻都市人的生活方式。

与 Keys 楼同侧的是 Verandah 楼（第 5、6、7 栋），它们公用的屋顶露台以及装饰有绿化植物的建筑立面和屋顶，都使得它们与众不同。这几栋住宅独特之处还在于：它们不仅拥有前面的阳台，另外在厨房边上还有一个花园露台，这给住户客厅和餐厅的布置提供了更多的选择。

第 8 栋楼由于地基最高，所以成为这些建筑群中最高的住宅楼，并可俯瞰迷人的水景。与其他 Verandah 楼一样，每个单位都有两个正面并且都是对流通风。

Hideaway 和 Pavilion 楼（第 9、10、11 栋）大多是四房或者四房以上的单位，并且配有宽敞隔热的客厅阳台和餐厅阳台以及与主卧相通的拐角阳台。Hideaway 楼（第 9、10 栋）在一楼设有 SPA 会所。Pavilion 楼（第 11 栋）对大泳池一览无遗，并且可以毫无阻挡地欣赏水景以及附近高档平房周边的绿化。

Location: 7 Ardmore Park, Singapore

Architectural Design: Ben van Berkel / UNStudio

Design Team: Ben van Berkel with Wouter de Jonge and Holger Hoffmann, Imola Berczi, Christian Bergmann, Aurelie Hsiao, Juergen Heinzel, Derrick Diporedjo, Nanang Santoso, Joerg Petri, Kristin Sandner, Katrin Zauner, Arne Nielsen

Facade Design: Ove Arup, Singapore

Interior Design Show Suite: Terry Hunziker

Floor Area: 15,666m^2 of apartments, plus 4,400 m^2 carpark

Land Area: 5,625m^2

Photography: Iwan Baan, Pontiac Land Group

项目地点：新加坡雅茂园 7 号

建筑设计：Ben van Berkel / UNStudio 建筑事务所

设计团队：Ben van Berkel, Wouter de Jonge、Holger Hoffmann、Imola Berczi、Christian Bergmann、Aurelie Hsiao、Juergen Heinzel、Derrick Diporedjo、Nanang Santoso、Joerg Petri、Kristin Sandner、Katrin Zauner、Arne Nielsen

外立面设计：Ove Arup，新加坡

室内样板间设计：Terry Hunziker

建筑面积：15 666 m^2（公寓）4 400 m^2（停车场）

占地面积：5 625 m^2

摄影：Iwan Baan、邦典置地公司

Keywords 关键词

Bay Windows 大玻璃窗

Double-height Balconies 双高阳台

Column-free Corners 无柱角落

Ardmore Residence

雅茂园高层公寓

Overview

The Ardmore Residence at 7 Ardmore Park in Singapore is located in a prime location close to the Orchard Road luxury shopping district and enjoys both expansive views of the panoramic cityscape of Singapore City and the vast green areas of its immediate western and eastern surroundings.

Design Concept

The primary concept for the design of the 36 storey, 17,178 m^2 residential tower is a multi-layered architectural response to the natural landscape inherent to the "Garden City" of Singapore. This landscape concept is integrated into the design by means of four large details: the articulation of the facade, which through its detailing creates various organic textures and patterns; expansive views across the city made possible by large glazed areas, bay windows and double-height balconies; the interior "living landscape" concept adopted for the design of the two apartment types and the introduction of transparency and connectivity to the ground level gardens by means of a raised structure supported by an open framework.

Facade Design

The facade of the Ardmore Residence is derived from micro-design features which interweave structural elements, such as bay windows and balconies into one continuous line. The facade pattern is repeated for every four storeys of the building, whilst rounded glass creates column-free corners, visually merging the internal spaces with the external balconies. Intertwining lines and surfaces wrap the apartments, seamlessly incorporating sun screening, whilst also ensuring that the inner qualities of the apartments and the outer appearance of the building together form a unified whole. From a distance the tower appears to adopt vastly divergent contours when viewed from different perspectives, whilst from close by the various openings in the concrete panels of the facade affect a sense of organic mutation and transition as you move around the building.

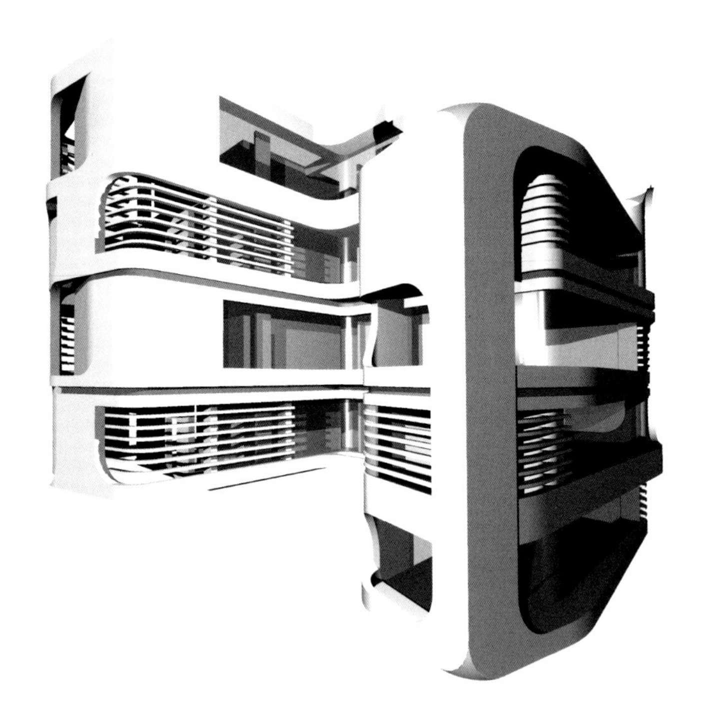

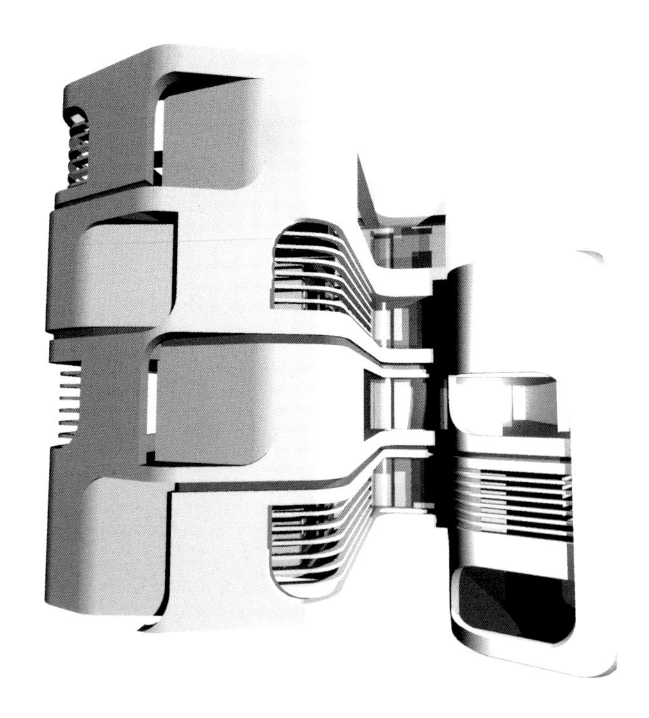

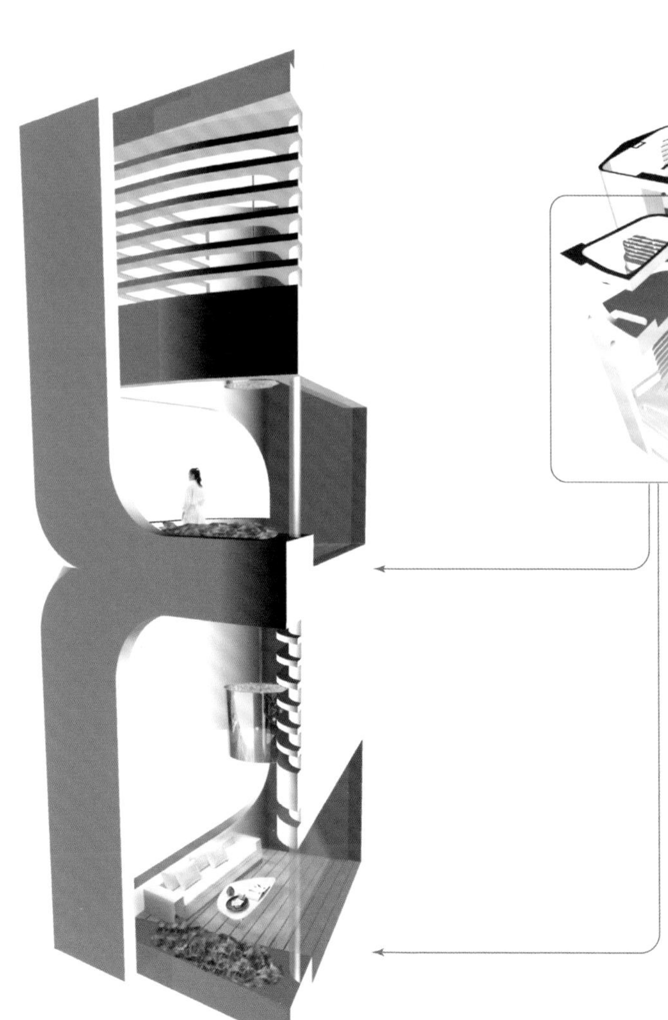

the facade system consists of
a number of highly repetitive
"frames" that range in scale from
one to four floor heights to
blur the boundaries of the horizontally
stacked floors visually

AND

by doing so provides unique spatial
qualities for the outdoor spaces:
"framing" the space, protecting you from
being seen and blocking the heavy
western sun.
like this, a human comfortable scale merges
with an outstanding outdoor experience

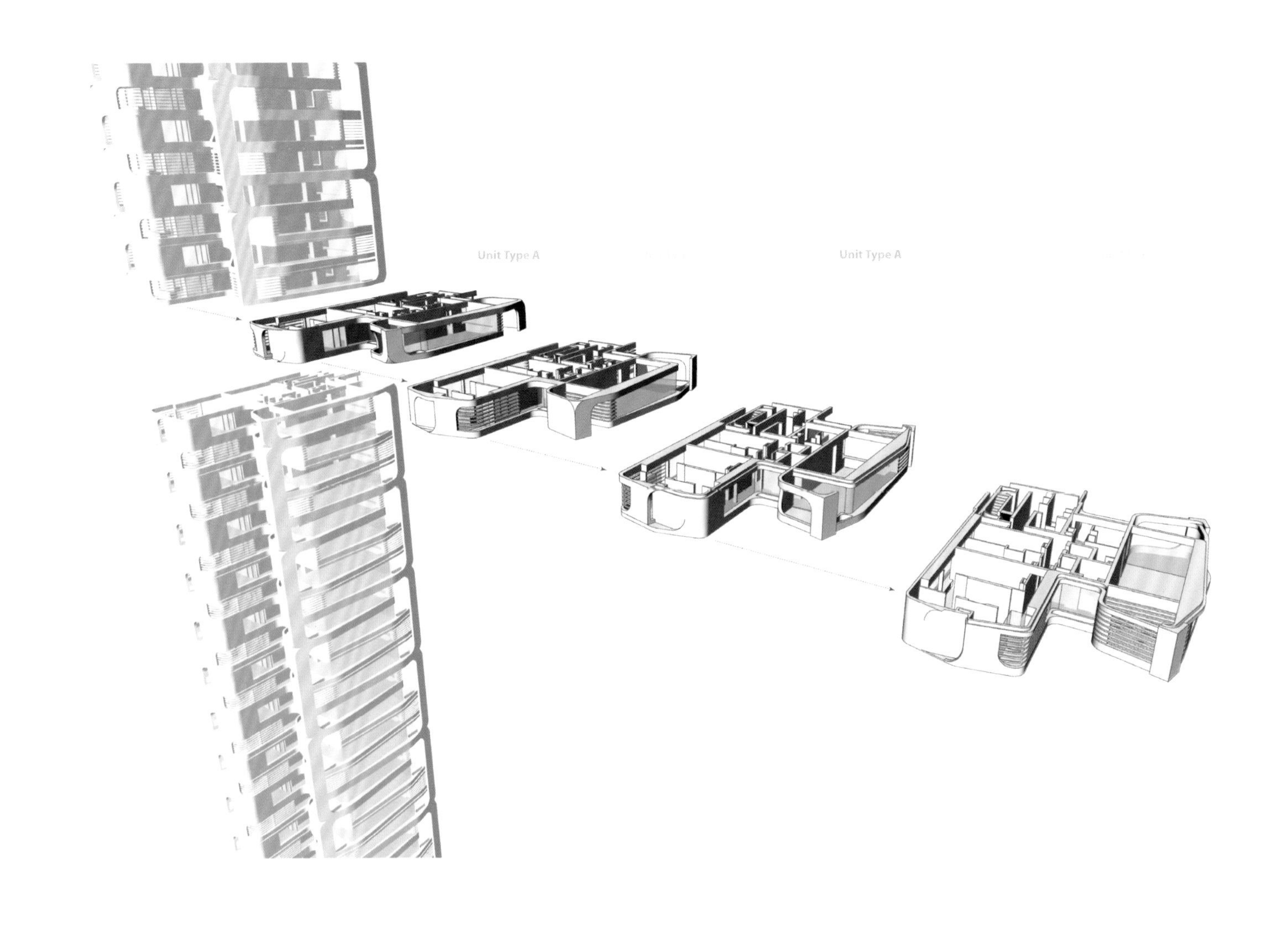
Unit Type A
Unit Type A

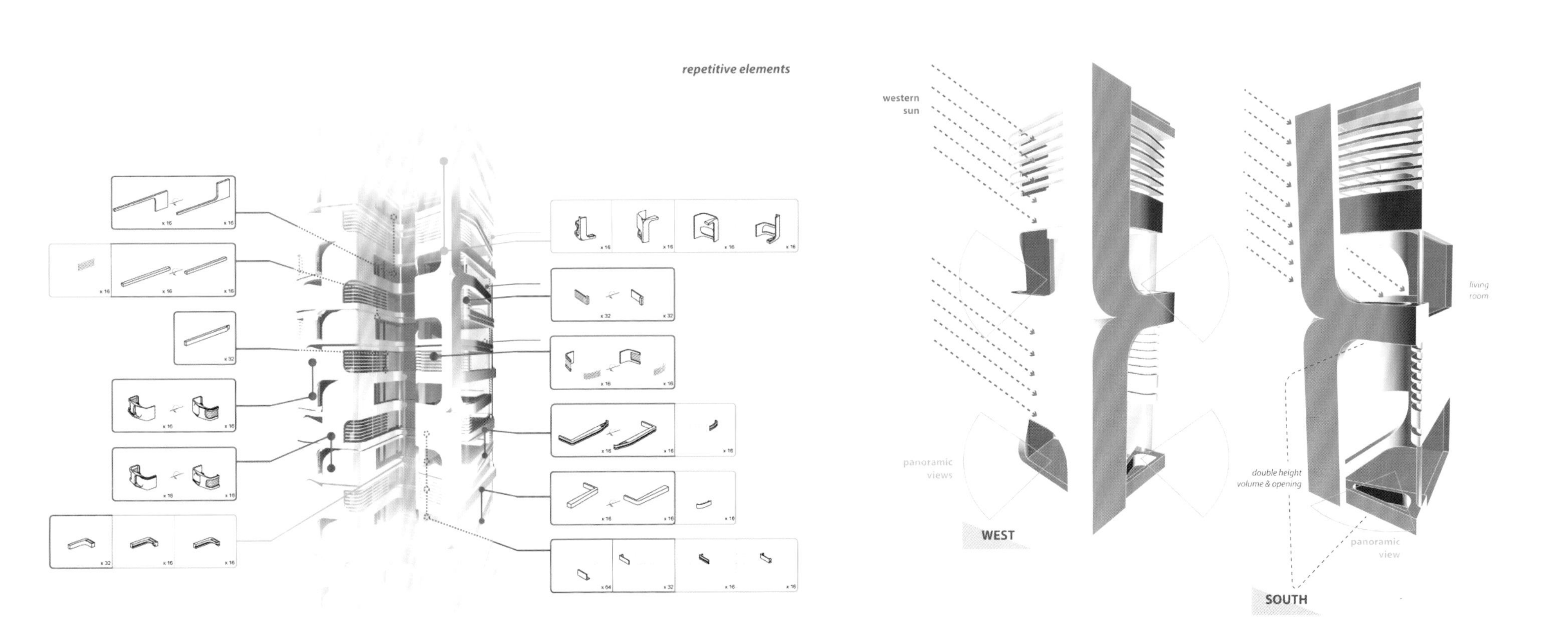
repetitive elements
western
sun
living
room
panoramic
views
double height
volume & opening
panoramic
view
WEST
SOUTH

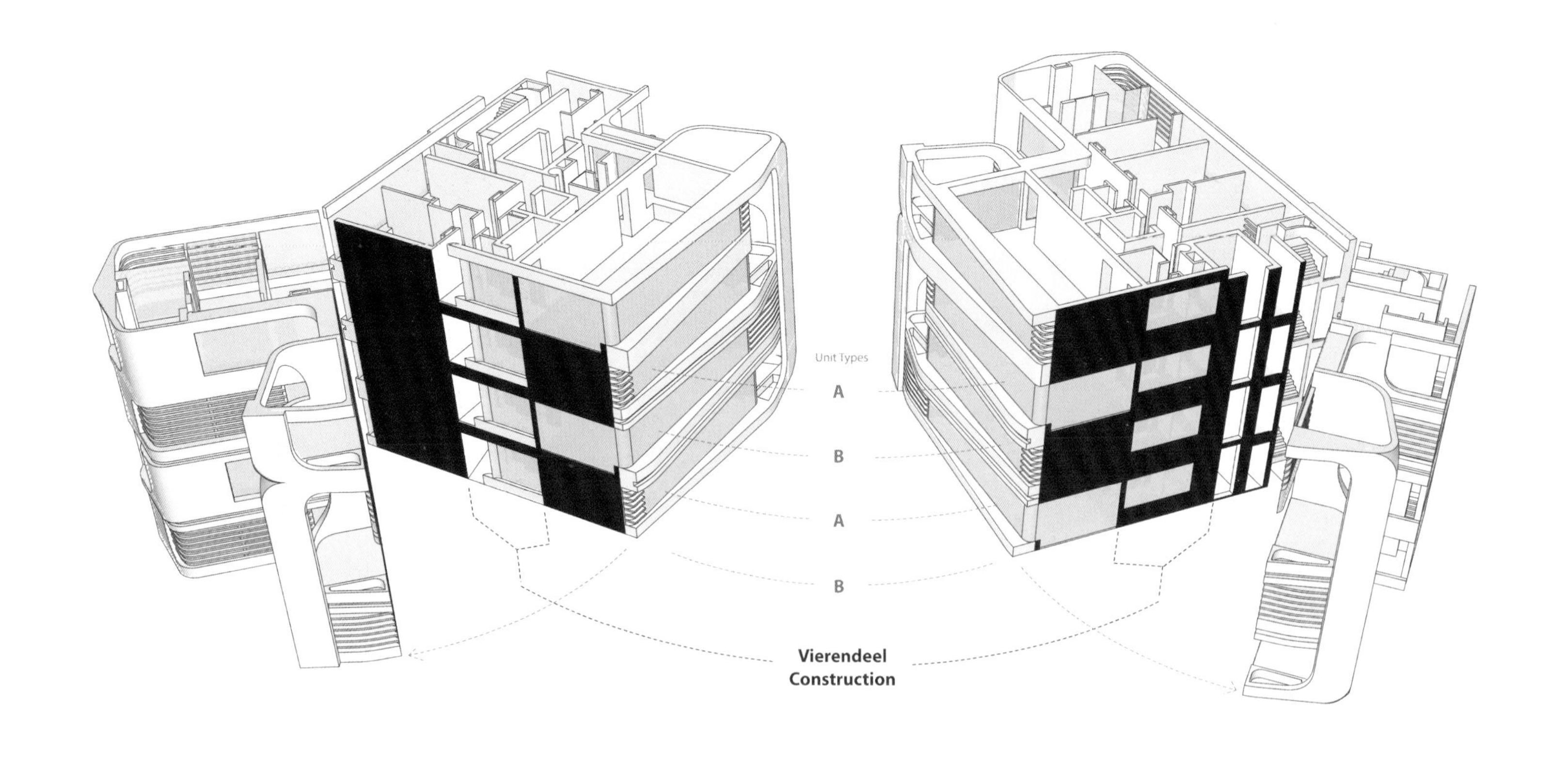
Unit Types
A
B
A
B
Vierendeel
Construction

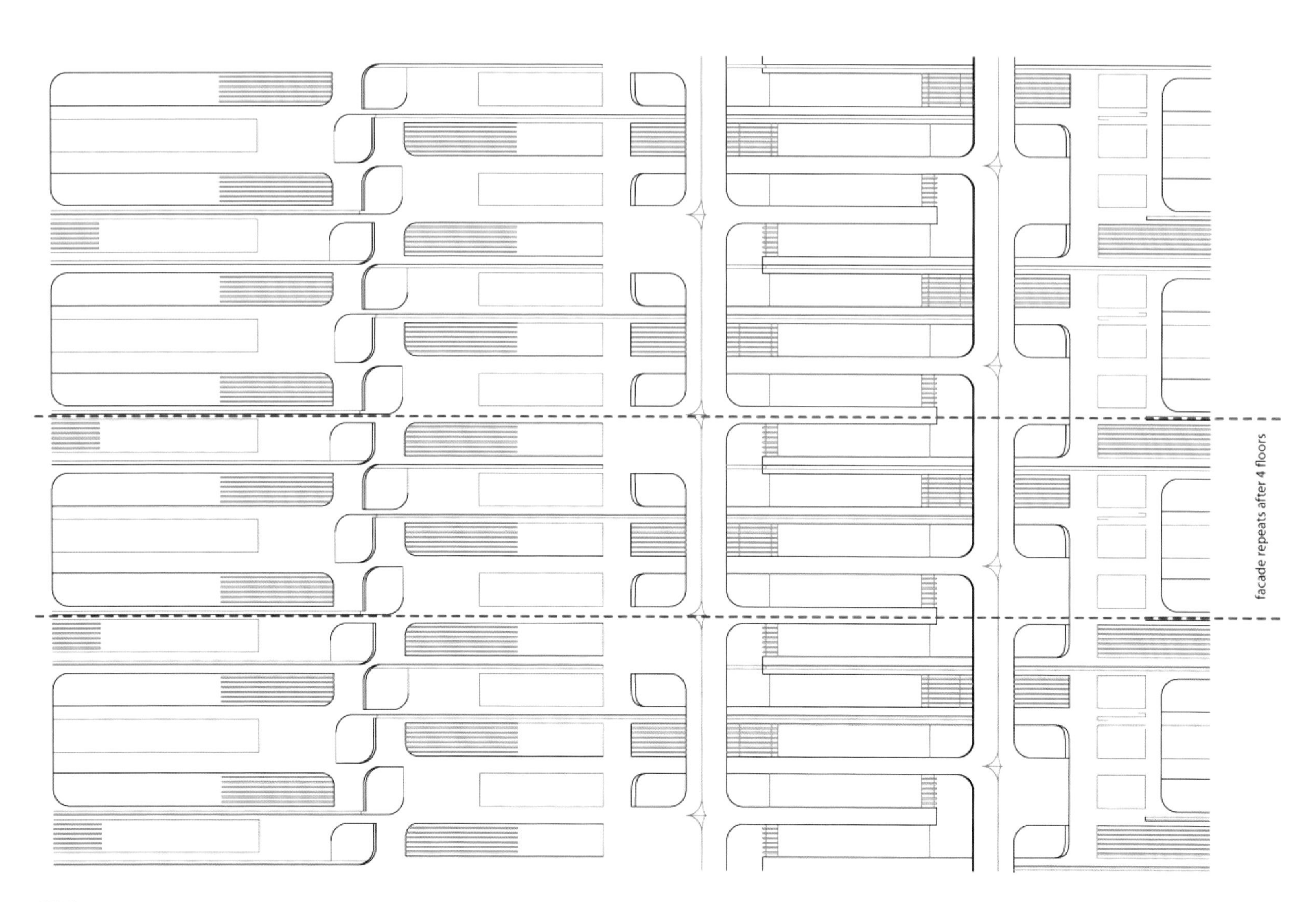
facade repeats after 4 floors

项目概况

由 UNstudio 的德国分部设计完成的雅茂公寓位于新加坡雅茂园 7 号，拥有优越的地理环境，紧邻乌节路豪华购物区，可俯瞰新加坡市容全景和周围东西两侧的宽阔绿化区。

设计理念

这栋 36 层高、17 178 m^2 的住宅大楼的主要设计理念是多层建筑与新加坡“花园城市”固有自然景观的相互呼应。设计师通过四大细节将这种景观概念融入设计：通过细节构成多样化的有机纹理和图案，打造连贯的建筑外观；通过大玻璃窗、飘窗和双高阳台，实现广阔的城市视野；通过采用“居住景观”的理念，呈现两种公寓类型的设计；以及通过一个开放的框架支撑起一个凸起的结构，实现地面花园的透明度和贯通性。

立面设计

雅茂园的立面设计是由细节设计发生的。这一设计将诸如凸窗和阳台等结构元素交织形成连续的线条。公寓每隔 4 层重复同样的立面外观，圆润的玻璃打造出无柱角落，在视觉上将室内空间和室外阳台相互融合。交织的线条和表面包裹着整栋公寓，在实现持续的日照屏蔽时，也确保公寓的内在品质和建筑物的外观形成一个统一的整体。

Site Plan 总平面图

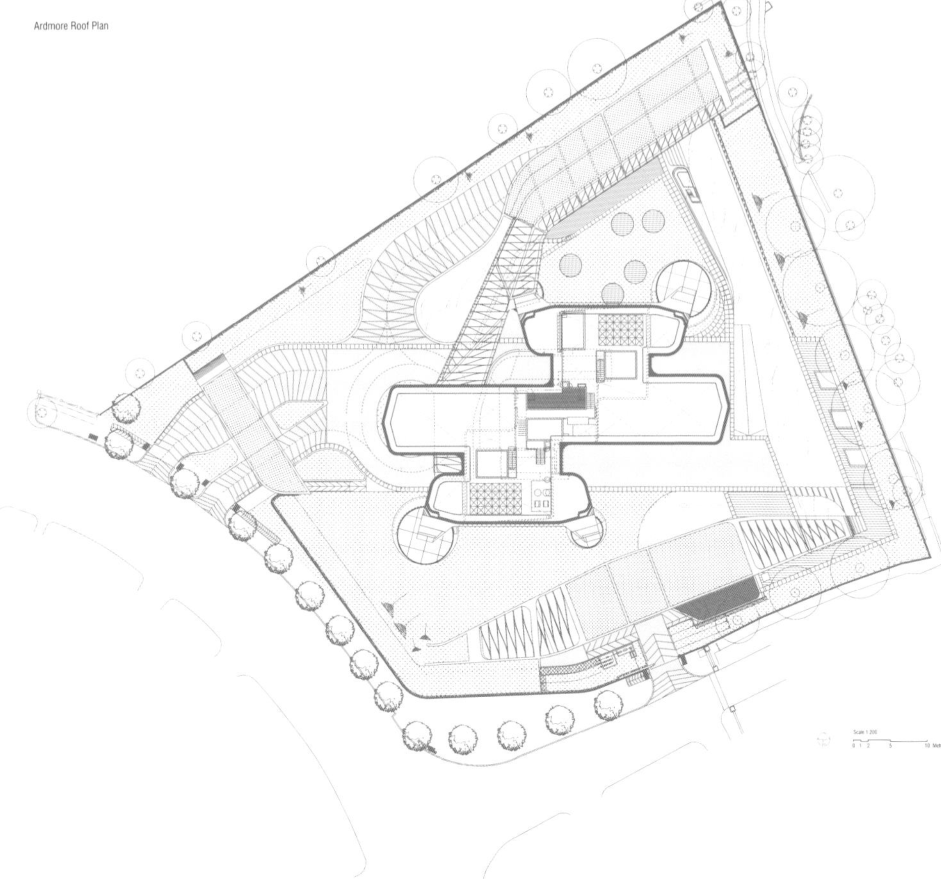

Roof Plan　屋顶层平面图

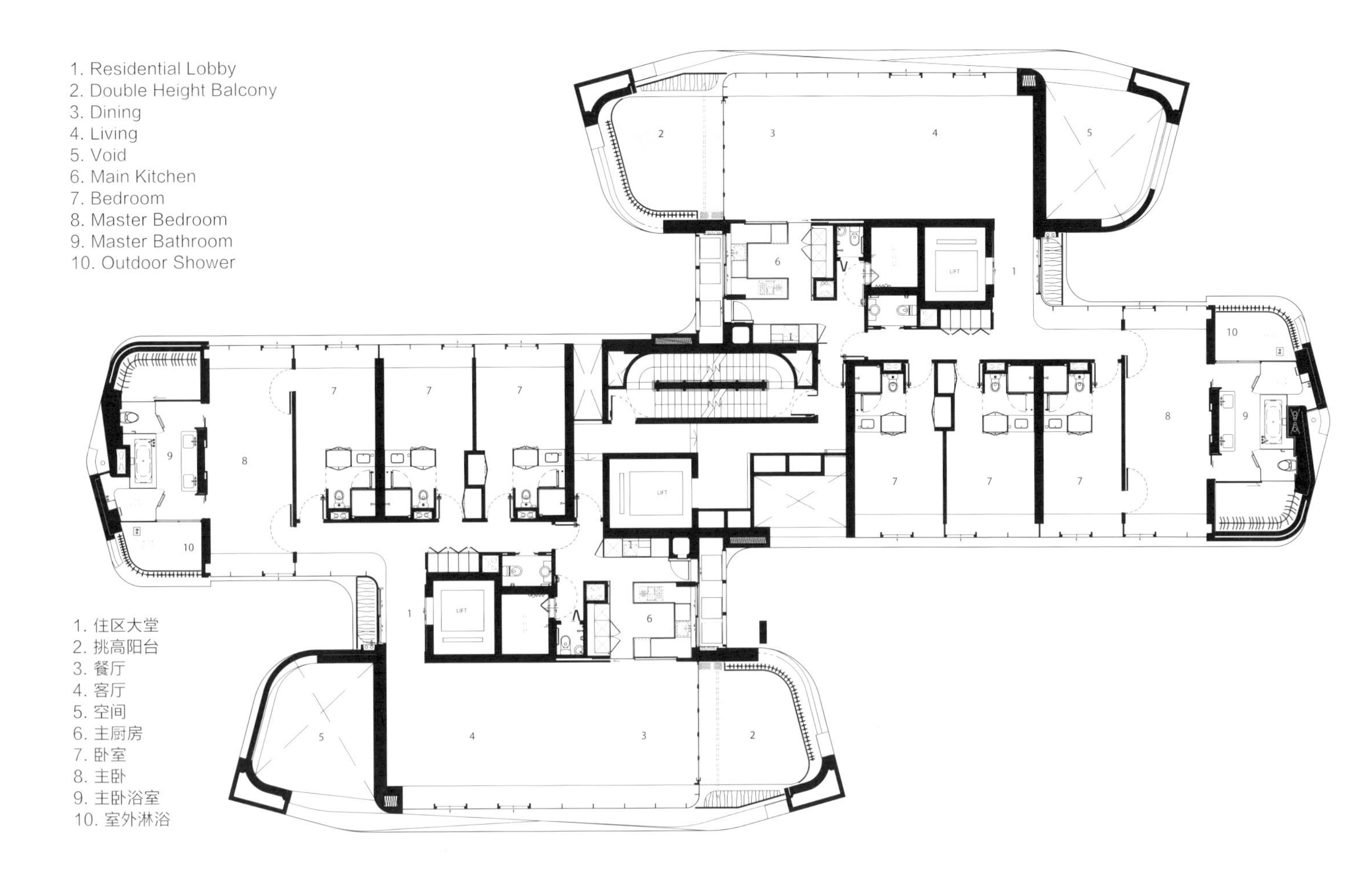

Floor Plan 08　八层平面图

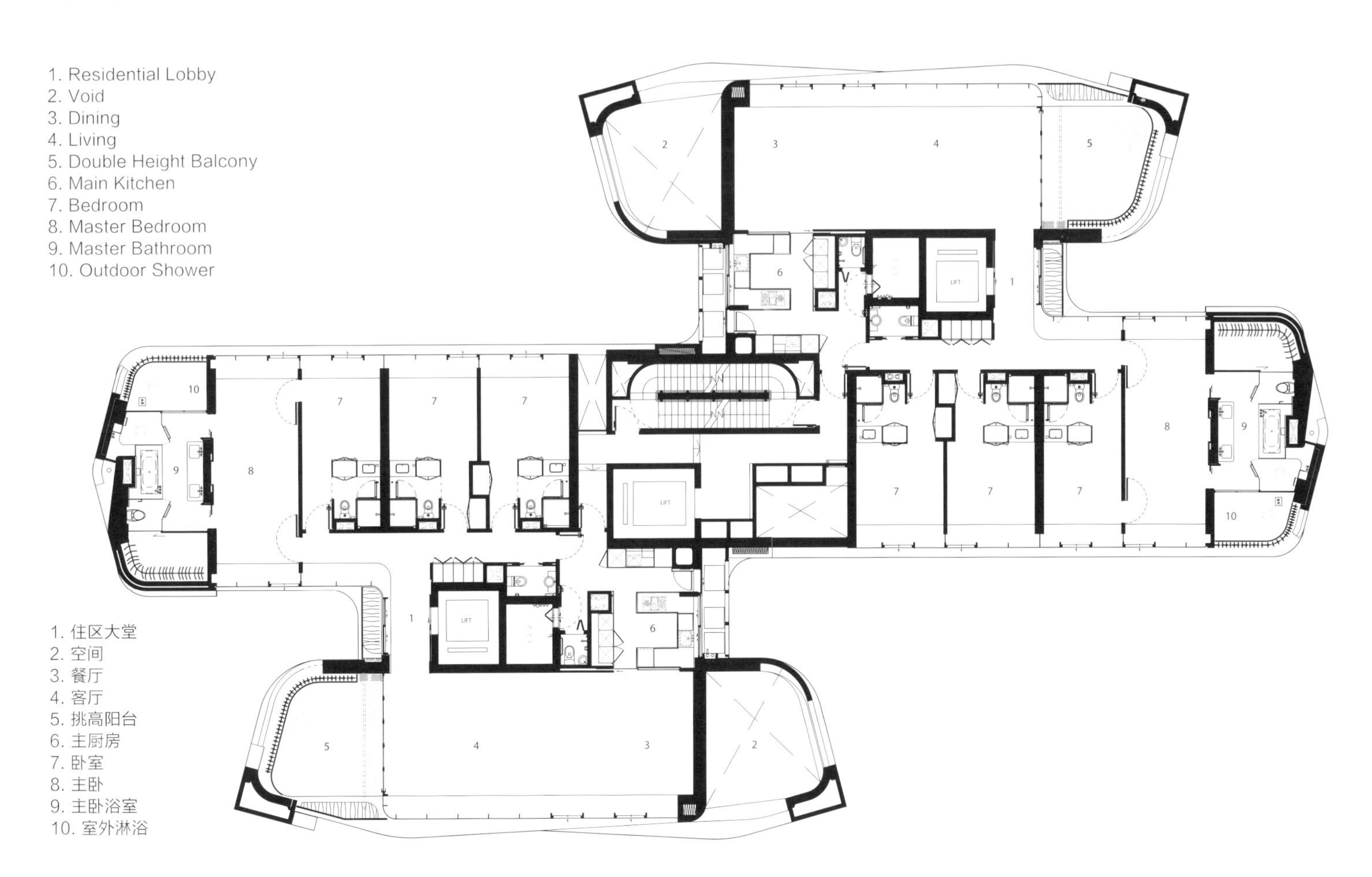

Floor Plan 09　九层平面图

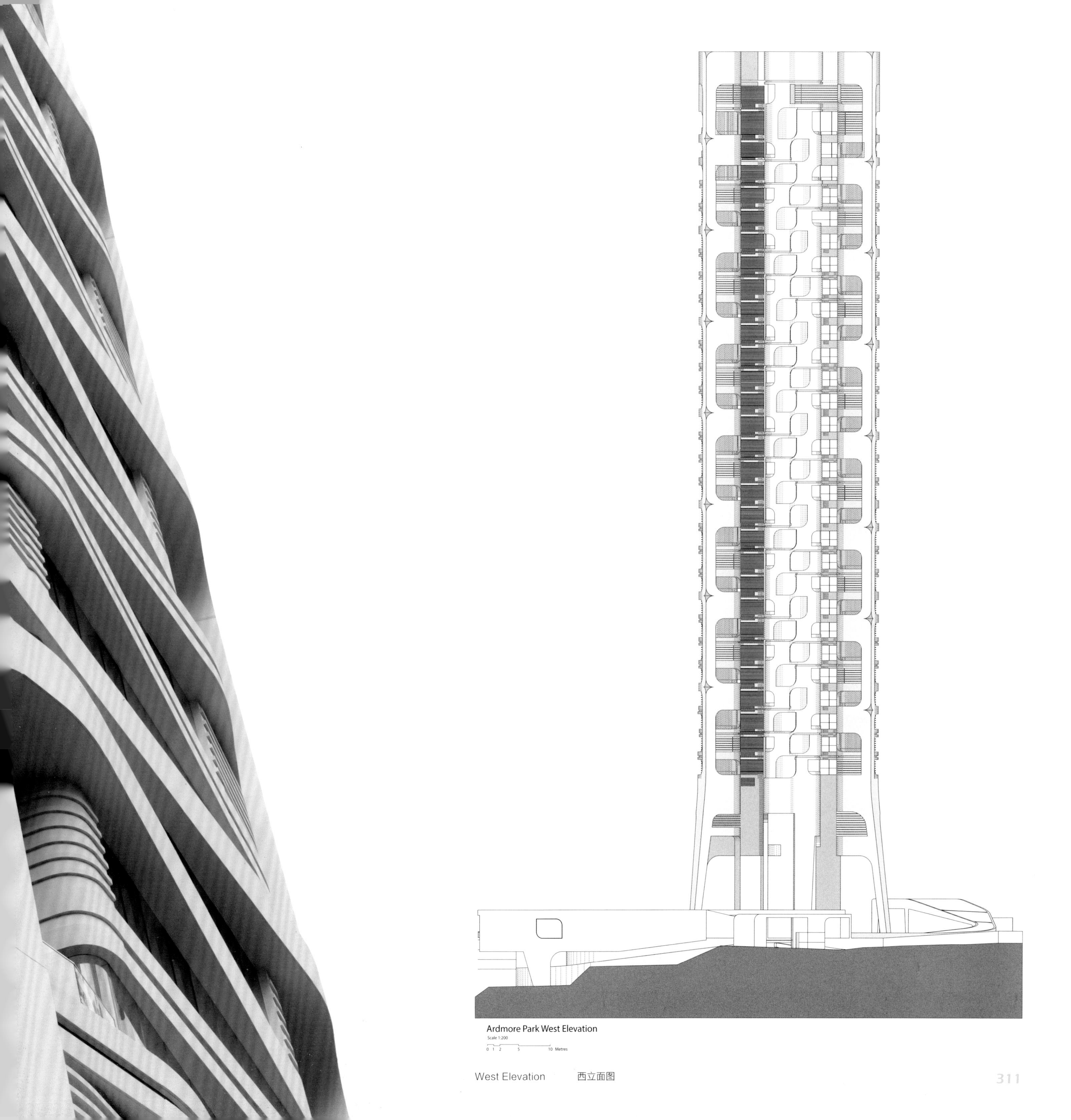

West Elevation 西立面图

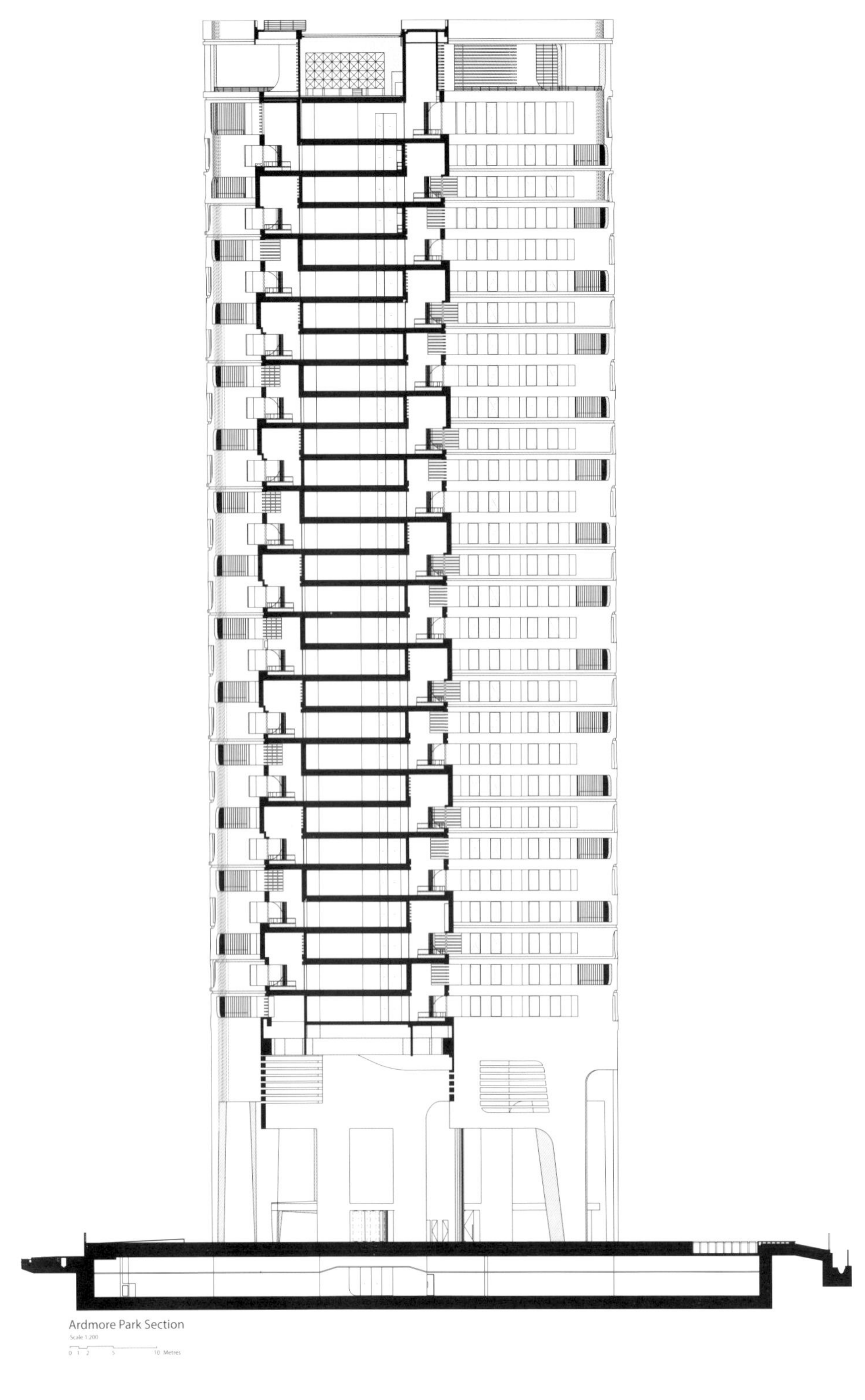

CC Section　　剖面图 CC

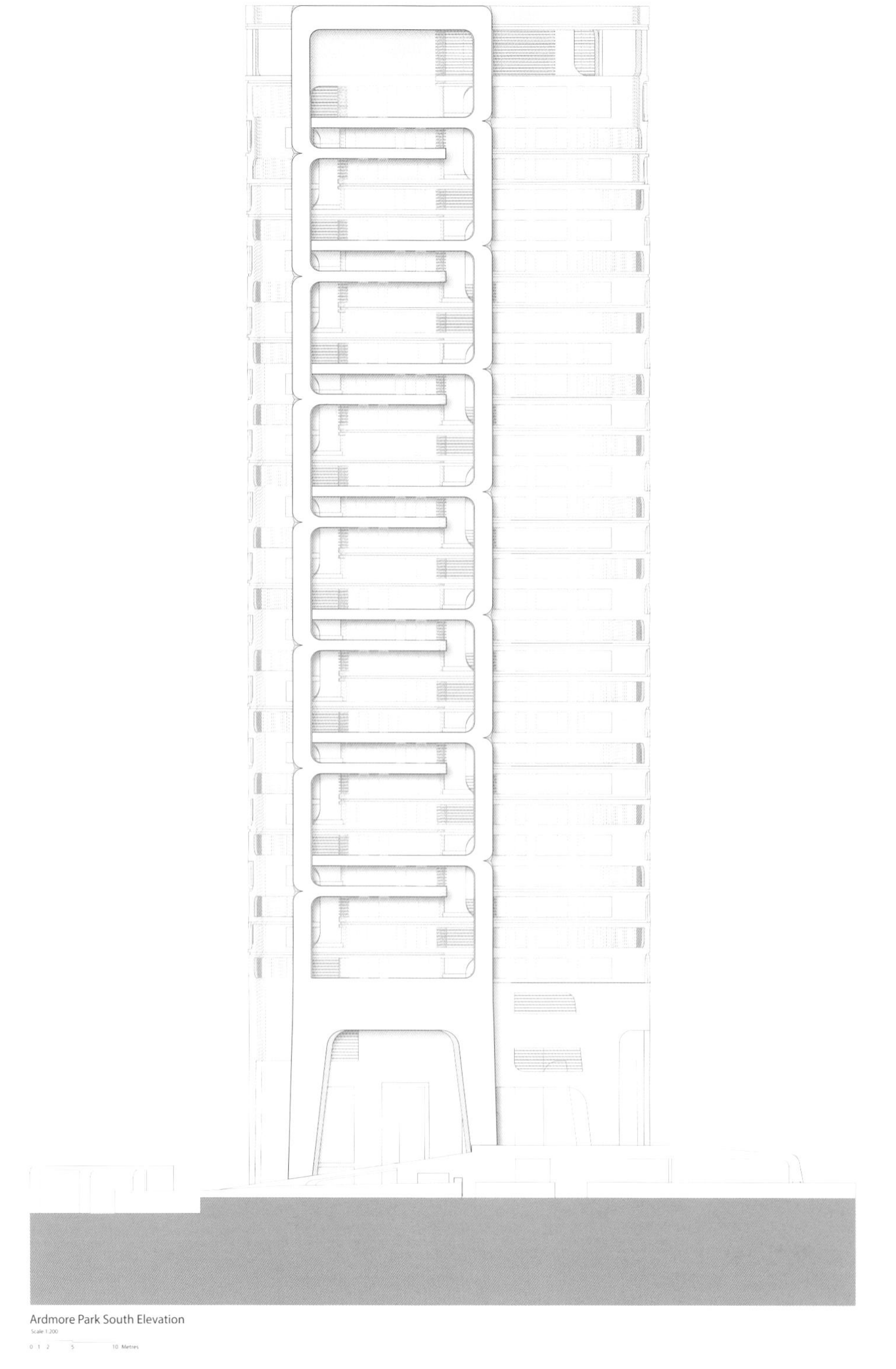

South Elevation　　南立面图

1. Fully retractable aluminium louvers as balcony screen
2. 500H tempered laminated glass screen on 500mm high RC parapet
3. 150mm stone flooring
4. 250mm RC slab
5. Rain water collection for semi outdoor shower at balcony for every floor to be channelled to sewerage system
6. 175mm RC slab
7. RC structural wall
8. GFRC panel to precast concrete facade cladding
9. 50mm timber flooring
10. Slidable and foldable glass door

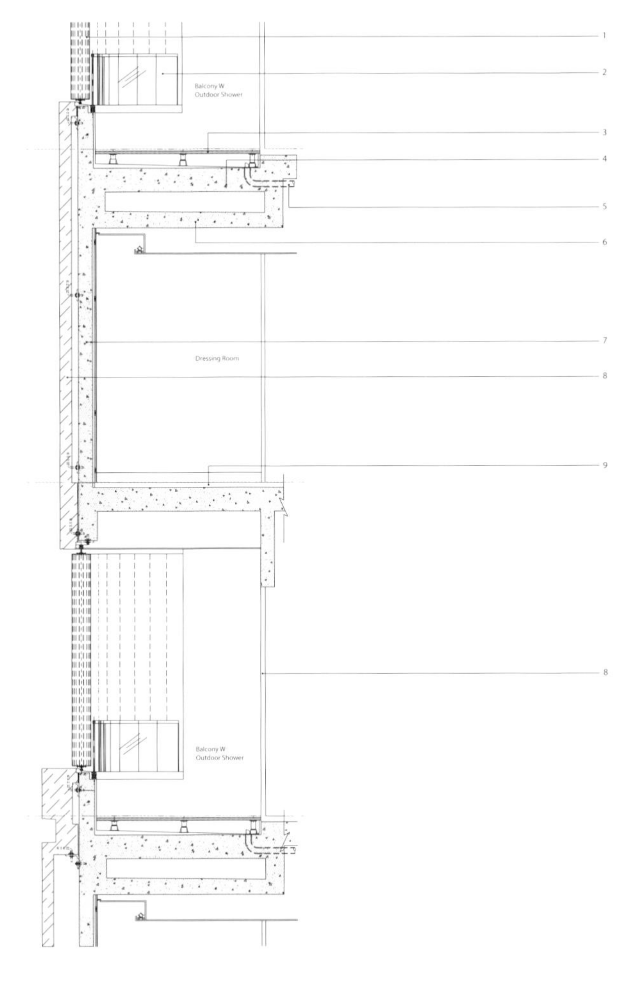

Outdoor Shower detail　室外淋浴细节

1. 阳台可伸缩铝制百叶屏风
2. 0.5 米高护栏用夹层钢化玻璃屏风
3. 15 厘米厚石材地板
4. 25 厘米厚栏板
5. 收集雨水用于每层阳台处的半室外淋浴，并导入排污系统
6.17.5 厘米厚栏板
7. 结构墙
8. 预制混凝土立面覆盖 GFRC 板
9. 5 厘米厚木质地板
10. 可滑动 / 折叠玻璃门

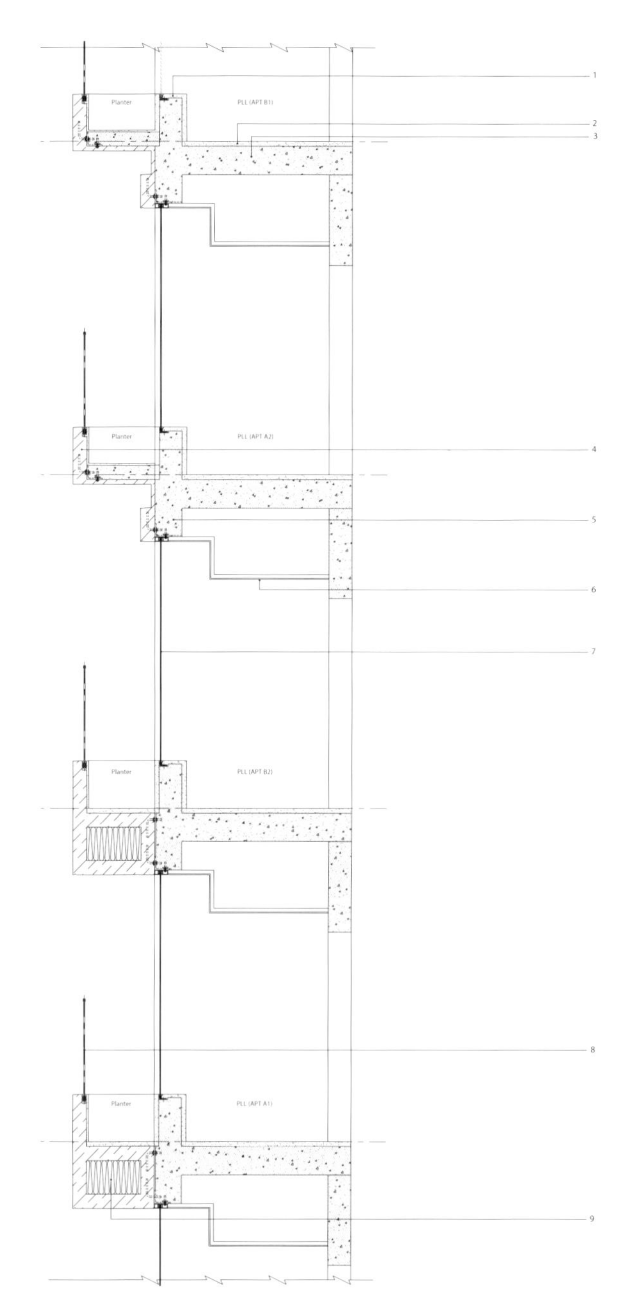

Planter detail　种植详图

1. White interior finish
2. 50mm Stone flooring
3. 300mm RC slab
4. GFRC panel to precast concrete facade cladding
5. 600 BY 300 RC beam
6. 30mm Ceiling surface
7. Laminated heat strengthened clear glass
8. 1000H Tempered laminated glass railing on 500H raised planter box
9. Styrofoam

1. 白色室内装修
2. 5 厘米厚石材地板
3. 30 厘米厚栏板
4. 预制混凝土立面覆盖 GFRC 板
5. 600*300 横梁
6. 3 厘米厚吊顶表面
7. 隔热、强化透明玻璃
8. 种植箱上的钢化玻璃栏杆
9. 泡沫聚苯乙烯

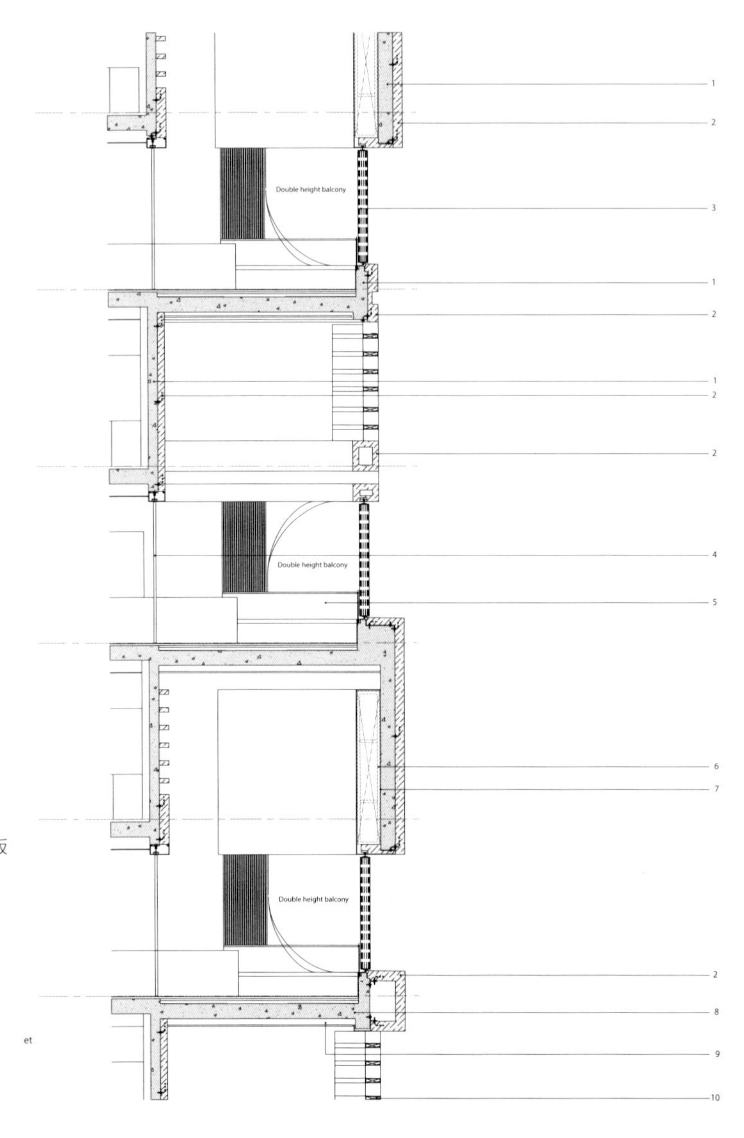

1. 结构墙
2. 预制混凝土立面 GFRC 覆盖板
3. 阳台可伸缩铝制百叶屏风
4. 可滑动 / 折叠夹层玻璃门
5. 0.5 米高护栏用的钢化玻璃
6. 木制表皮
7. 钢骨架
8. 30 厘米厚拦板
9. 12 厘米机电
10. 铝制屏风

Balcony detail　阳台细节图

1. RC stuctural wall
2. GFRC panel to precast concrete facade cladding
3. Fully retractable aluminium louvers as balcony screen
4. Sliding and folding laminated glass door
5. 500H tempered laminated glass screen on 500mm high RC parapet
6. Wooden cladding
7. Steel frame
8. 300mm RC slab
9. 120mm M&E
10. Aluminium screen